621.319 D754i   FV
DOYLE
AN INTRODUCTION TO ELECTRICAL
WIRING
                        14.00

JUNIOR COLLEGE DISTRICT
of St. Louis - St. Louis County
LIBRARY
5801 Wilson Ave.
St. Louis, Missouri 63110

An Introduction to Electrical Wiring

An Introduction to Electrical Wiring

John M. Doyle
Director
Educational Research and Development
International Correspondence Schools

RESTON PUBLISHING COMPANY, INC.
Reston, Virginia 22090
A Prentice-Hall Company

Library of Congress Cataloging in Publication Data

Doyle, John M
    An introduction to electrical wiring.

    1. Electric wiring, Interior.   I. Title.
TK3271.D69   621.319'24   75-1455
ISBN 0-87909-369-2

© 1975 by
Reston Publishing Company, Inc.
A Prentice-Hall Company
Reston, Virginia 22090

10   9   8   7   6   5   4   3   2   1

To My Daughter Tamara Elaine

Contents

Part 2   *89*

4   Electrical Conductors   91

5   Circuit Components   115

Part 3   *131*

6   System and Equipment Grounding   133

7   Wiring Methods   151

Preface

This book is written for electrician trainees in vocational-technical schools, technical-institutes, and industrial apprenticeship programs.

In keeping with its title, Introduction to Electrical Wiring, attention is centered on single-family dwellings, both urban and rural. Once the reader has acquired experience in the wiring of such structures, the changeover to commercial and/or industrial systems should proceed smoothly.

This is an entirely practical text. The only mathematics used—and sparingly—are arithmetic and the most elementary forms of algebra.

All methods discussed are in accordance with the 1975 edition of the National Electrical Code (NEC), published by the National Fire Protection Association, 470 Atlantic Avenue, Boston, Mass. 02210. All students, trainees, and apprentices should have a copy or be able to easily refer to one.

The book is divided into five parts. Part 1 is a simplified but entirely adequate treatment of the basic concepts of electricity, magnetism, and illumination. Part 2 deals with the sizing, characteristics, and permissible uses of various types of electrical conductors and introduces the reader to common circuit components, such as switches, receptacles, and fuses. Part 3 covers the general considerations involved in any wiring job, such as wiring methods, grounding, the sizing of service entrances, and the provision of adequate lighting. Part 4 gives detailed instructions for the installation of complete wiring systems in buildings either under construction (new work) or already constructed (old work). Part 5 covers the various types of motors and methods for farm installations.

Because actual hardware is handled in any worthwhile training program, photographs of such hardware are not included in this book. Illustrations are used liberally, however, including the pictorial type, where such usage clarifies the discussion.

The reader may note some repetition in the book. This is intentional and is done to reinforce certain basics that the electrician must remember at all times.

Practical examples are scattered throughout the book and questions and/or problems appear at the end of each chapter. In addition, numerous tables, reprinted either in whole or in part by permission of the National Fire Protection Association, are included; the reader should also make use of the more extensive data given in the NEC.

I am grateful for the assistance of the NFPA. I would also like to express my appreciation to my colleagues at the International Correspondence Schools for their reviews of the manuscript and their many helpful comments. Particular thanks go to Howard Miller, P.E., Director, Electrical-Engineering School; Stewart Crouse, P.E., Director, Electrician's School; and Wallace Cullen, P.E., Director of Education, Votech Schools.

Finally, I am most grateful for my wife's assistance. After many years of manuscript typing, proofreading, and indexing, she is still enthusiastic about each new book I undertake. In a very real sense, every book is also dedicated to her.

John M. Doyle

Part
1

1
Electricity and Magnetism

1.1   *Electric Charges*

When two clean, small glass balls are suspended near each other by separate threads, they hang straight down next to each other. See Fig. 1-1(a). If the balls are rubbed with a silk cloth, they no longer hang straight down, but *repel* each other, as shown in Fig. 1-1(b).

Several important deductions can be made from this experiment. (1) rubbing the balls with a silk cloth produces a *force of repulsion* between the balls. This force is called an *electric charge*. (2) When the balls are uncharged, no force exists between them; and they are said to be *electrically neutral*. (3) Since both balls are rubbed with the same material, they must acquire the *same kind* of charge. (4) That is, *like charges repel*.

Let a small rubber ball replace one glass ball. When both balls are again rubbed with a silk cloth, they move toward each other as shown in Fig. 1-1(c). What does this indicate? Again the experiment leads to important deductions. (1) The rubber ball must acquire a *different kind* of charge than the glass ball. (2) That *unlike charges attract*.

The two opposite kinds of electric charge are called *positive* and *negative*. The condition of positiveness or negativeness is called *polarity*. Positive polarity is indicated by a plus sign (+) and negative by a minus sign (−).

3

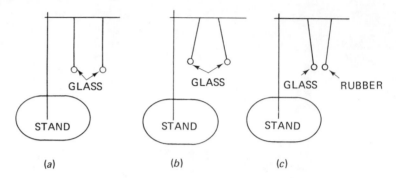

GLASS   GLASS   GLASS   RUBBER

STAND   STAND   STAND

(a)   (b)   (c)

Figure 1-1   Demonstrating the laws of attraction and repulsion between electric charges: (a) electrically neutral, (b) like charges repel, and (c) unlike charges attract

Because the form of electricity demonstrated by the above experiments is motionless, it is called *static* electricity.

Finally, it should be noted that a charged (electrified) body retains its charge for a certain length of time, but eventually the electrification leaks off. If the body comes into contact with earth (ground) it discharges quickly.

Now that you know the meaning of charge (electrification), the two kinds of charges, and the laws of attraction and repulsion between charges, you are ready to study the *electron theory* which is so named because it explains how electric energy in the form of electrons moves from one point to another in an electric circuit.

1.2   *The Electron Theory*

The invisible and smallest whole elemental component called the *atom*, is the basic building block of all matter whether solid, liquid, or gas. It is often pictured as shown in Fig. 1-2. The central portion, called the *nucleus*, has a positive electric charge. Spinning around that nucleus in egg-shaped orbits are elementary particles called *electrons*. Each electron has a negative charge; and the sum of these negative charges, under normal conditions, is exactly equal to the positive charge of the nucleus. Since the two charges are equal to one another, the *net* electrical charge of a *normal atom* (one containing a specified stable number of electrons) is zero.

Because they have unlike charges, a force of attraction exists between the nucleus and its orbiting electrons. This force is exactly balanced, however, by a *centrifugal force* that tends to push the electrons outward from the center of rotation. As a result, the electrons assume *fixed orbits* around their nucleus.

For instance, there are 29 electrons in one atom of copper. Of these, 28 are bound permanently—at least, for our purposes—to the nucleus, but only a very weak binding force exists between the nucleus and the one remaining

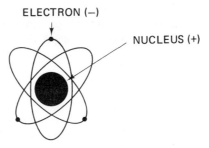

Figure 1-2   Basic structure of an atom

electron. This twenty-ninth electron, located in the outermost orbit, is called a *valence* electron.

When a valence electron finds itself midway between two nuclei, as shown in Fig. 1-3, it is attracted equally by each nucleus. In this position of balanced attraction, a valence electron often leaves its own orbit and moves into the outermost orbit of a neighboring atom. For this reason valence electrons are often called *free* electrons. *The movement of free electrons is the means by which electrical energy is carried from one point to another.*

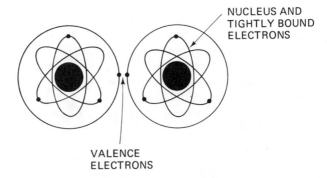

Figure 1-3   When a valence electron finds itself midway between two nuclei, it is attracted equally by each nucleus

1.3   *Conductors and Insulators*

When a material contains many free electrons *per unit volume* (for example, per cubic inch) it is called a *conductor*. Although any metal may be used as an electric conductor, copper is the most frequent choice. Aluminum is another very popular conductor.

In electric circuits it is often necessary to confine the free electrons to a given path or to prevent contact either *between* or *with* conductors. Materials containing very few free electrons per unit volume are used for this purpose, and are called *insulators*. Plastic, rubber, porcelain, glass, and air are the most common insulators encountered in electrical wiring.

The operation of most switches depends on the insulating properties of air. Simply by opening (breaking) the conducting path at a selected point, a switch prevents the transfer of free electrons and, therefore, the transfer of electric energy from one point in a conductor to another.

When the electric *pressure* (see Sec. 1-6) applied across an insulator is greater than that specified by the manufacturer, the insulator breaks down and becomes a conductor. The arcing of electricity across an open switch is one such example of insulator breakdown. For this reason, insulators must always be operated within their specified ratings.

1.4   *The Parts of an Electrical Circuit*

All practical electric circuits, regardless of how simple or how complex they may be, require four basic parts: (1) a *source* of electrical energy that forces free electrons to flow through the circuit; (2) *conductors* for carrying the free electrons around the complete circuit; (3) a *load*, that is, a device or devices, to which electrical energy is supplied; and (4) a *control device* to turn the circuit "on" or "off."

The *schematic* diagram of a simple electric circuit is shown in Fig. 1-4. The source may be an electric outlet in a home, a battery, an electric generator, or some other device. As you will learn later, either of two basic types of sources are used—*alternating current* (ac) or *direct current* (dc). The symbol shown in Fig. 1-4 is for a direct-current source. A different symbol is used to indicate an alternating-current source.

The conductors are usually copper wire; however, aluminum may also be used.

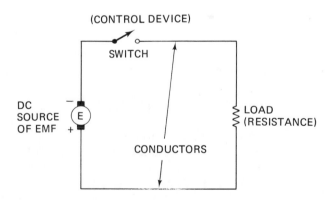

Figure 1-4   The schematic diagram of a simple electric circuit

The control device in Fig. 1-4 is a *switch*. When the switch is open, as shown, free electrons cannot move around the circuit. A complete path which permits the flow of free electrons is formed when the switch is closed.

The load may be any one of a wide variety of devices, such as a light bulb, a motor, a washer, or a dryer. The symbol for a *resistance* (a device which changes electrical energy into heat energy) is also shown in Fig. 1-4. As you will learn, different schematic symbols are used to represent different kinds of loads.

1.5   *Current*

In working with electrical circuits, it is necessary to know the *rate-of-flow* of free electrons through the circuit; that is, how many free electrons pass a given point in the circuit in one second.

The rate-of-flow of free electrons is called *current* and is designated by the letter symbol *I* which indicates the *intensity* of electron flow. When a certain very large number of electrons (6.24×10^{18} to be exact) flow past a given point in one second, the rate of current flow is one *ampere*, abbreviated A. Hence, the *unit* of electric current is the ampere.

Prefixes

Very often, the basic unit of an electric quantity is either too small or too large for practical use. This difficulty is overcome by the use of *prefixes* which indicate either multiples or fractions of the basic units. The prefixes an electrician encounters frequently are shown in Table 1-1.

Prefix	Meaning	Abbreviation
Mega	One Million	M
Kilo	One thousand	k
Milli	One thousandth	m
Micro	One millionth	μ

Table 1-1   Common prefixes

Measuring current

Aside from the fact that electrons are invisible, it would be impossible to count them manually as they flowed by a given point in a circuit. Fortunately, an electrical instrument called either an *ammeter, milliammeter,* or *microammeter,* depending on its *range,* indicates directly the amount of current passing through a circuit.

The scale of a typical ammeter is *calibrated* (marked) as shown in Fig. 1-5. The maximum current-carrying capability of this particular meter is 1 A, the highest value indicated on the scale. Each major calibration mark represents 0.1 A (100 mA).

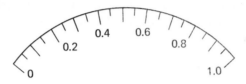

Figure 1-5   The scale of a typical ammeter

Usually, several sets of calibration marks are shown on the same scale and the appropriate range is selected either by means of a so-called *range switch* or by inserting plugs into appropriately marked *jacks*. Always read the manufacturer's literature that comes with any instrument before you try to use it.

Since an ammeter measures the current passing *through* a circuit, it is connected *end-to-end* with other components, as shown in Fig. 1-6. Notice that the schematic symbol for an ammeter is a circle containing the letter symbol A. For a milliammeter the correct symbol is mA encircled and for a microammeter μA encircled.

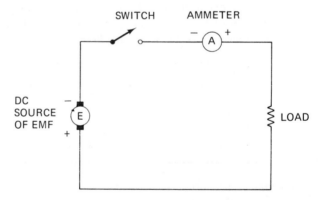

Figure 1-6   Correct method of connecting an ammeter in a dc circuit

When an ammeter is used to measure dc *it is necessary to observe polarity*. This means that the negative terminal of the meter must always be connected to a point of negative potential *with respect to* the positive terminal of the meter. If the meter is connected improperly, it can be ruined quickly.

When an ammeter is used to measure ac it is *not* necessary to observe polarity.

Finally, whenever you use a current-measuring instrument, *always* make sure to use a *range* that is large enough to measure the circuit current. If the probable value of circuit current is unknown, use the highest range available for the initial measurement. Then, if the current is small enough you can always switch to a lower range.

1.6   *Electric Pressure*

Whenever a source of electric energy is connected across the terminals of a *complete* electric circuit, it creates a *surplus* of free electrons at one terminal and a *deficiency* of free electrons at the other terminal. The terminal having the surplus of free electrons has a *net negative charge* and is indicated by a minus (−) sign. The terminal having the deficiency of free electrons has a *net positive charge* and is indicated by a plus (+) sign.

At the terminal charged positively, the free electrons are spaced further apart then normal and the forces of repulsion that act between them ("like charges repel") are reduced. This force of repulsion is a form of *potential energy* and is also called the *energy of position.* Let us see what the term potential energy means.

Energy is often defined as the ability (capacity) to do work. To illustrate: when a hammer is raised above a nail, the hammer, as a result of its position, has the potential ability to do work; that is, it possesses *potential* energy. As the hammer drops, it gains *kinetic* energy, the energy associated with movement. Finally, when the hammer strikes the nail, its energy is expended (dissipated) and work is accomplished—the nail is driven into the surface on which its point had rested.

The electrons in a conductor possess potential energy in much the same manner as the hammer; and work is done when the electrons repel (push) one another to a new position in the conductor.

From the above discussion, it is apparent that the potential energy of free electrons at the positive terminal of a circuit is *less* than the potential energy of those at the negative terminal. Thus, a *potential energy difference,* usually shortened to *potential difference,* exists across the terminals.

Whenever a potential difference exists between two points in a metallic conductor, the free electrons try to distribute themselves uniformly throughout the conductor and, in so doing, flow from the point of negative potential toward the point of positive potential; that is, from − to +. Thus *a potential difference is the pressure that creates current in an electric circuit.*

Because the source creates the potential difference that forces current through a circuit, it is called a source of *electromotive* or electron moving *force.* The usual abbreviation is EMF.

Now that we know both what is meant by the expression "source of EMF" and, that in a metallic conductor free electrons move from negative to positive, we can give a precise definition for current. *Current is the movement of free electrons through a circuit under the influence of an applied EMF.*

In some cases, the terminals of the source always maintain the same polarity and current flows in the same direction at all times through the circuit external to the source. Sources of this type are called *direct-current sources,* and the current they create is called *direct current,* abbreviated dc. In other cases, the terminals of the source reverse polarity periodically. Sources of this type are called *alternating-current sources,* and the current they create is called *alternating current,* abbreviated ac. For the remainder of this chapter, however, our only concern is with dc. Chapter 2 considers ac in detail.

In Fig. 1-7, we see that a dc source of EMF and a switch (control device) and a lamp (load) are connected by conductors. Since the source maintains a potential difference, it may be thought of as a *potential rise.* Moreover, since the electric energy provided by the source is used up in doing work, such as lighting the lamp in Fig. 1-7, the load represents a *potential drop.*

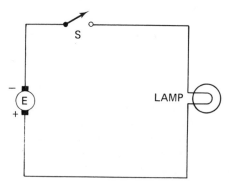

Figure 1-7   Simple electric circuit

In Sec. 1.12 and 1.13 you will see that the potential drops around an electric circuit are always equal to the potential rise; therefore, we use the same unit of measurement to represent either a potential rise or a potential drop. *The volt is the basic unit of potential difference.*

The letter symbol E is used to represent the potential rise occurring at the source while the letter symbol used to represent the potential drop occurring across a load is V. Both E and V are, of course, given in volts.

When the volt is either too small or too large for practical use, we can employ any of the prefixes introduced in Sec. 1-5.

Many electricians do not distinguish between "potential rise" and "potential drop" and refer to both as *voltage.* This interchangeable usage may be fine later on, but while you are using this text, please keep the distinction in mind.

Measuring potential differences

A *voltmeter* is used to measure either a potential voltage rise, E, or a voltage drop, V, across the terminals of a device. Thus a voltmeter is always connected *across* a device, as shown in Fig. 1-8. In this case, the voltmeter is being used to measure the voltage drop across one of several devices making up the total load.

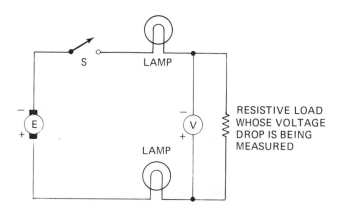

Figure 1-8   Correct method of connecting a voltmeter into a dc circuit

Since this is a dc circuit, polarity must be observed in connecting the meter; that is, − to − and + to +. If the voltmeter is connected in the wrong polarity, the meter pointer will move off scale and the meter may be damaged. It is also necessary of course, to use a range suitable for the voltage rise or drop being measured. The correct symbol for the voltmeter is a circle enclosing the letter V.

1.7   *Resistance*

As free electrons pick up velocity in their movement along the conductor, potential energy from the source is transferred to them in *kinetic* form; that is, the electrons acquire kinetic energy (the energy of motion). Before the electron has gone very far, however, it collides with a relatively large metal *ion*. An ion is simply an atom or group of atoms, which by loss or gain of one or more free electrons has acquired an electric charge. The ions take up fixed positions and give the metal conductor its shape. As a result of the collisions between free electrons and ions, the free electrons give up some of their kinetic energy to the ion in the form of *heat* energy. That is why a conductor feels warm to the touch when it is passing current. Since the heat energy released by the collisions into the surrounding air does not perform useful work, it is *wasted* energy. More important, if the amount of heat produced in the conductor is more than it can dissipate, the temperature of that conductor will rise and create a fire hazard. Many buildings have been destroyed as a result of such overheated conductors.

In passing from one point to another in an electric circuit, a given free electron is involved in many collisions. Since current is the movement of free electrons, the collisions oppose current. Another word for oppose is *resist;* accordingly, we call the opposition to current *resistance.* Stated formally, *resistance is that property of an electric circuit which opposes current.*

The *unit* of resistance is the *ohm,* and the unit symbol is the Greek letter Ω (omega). The letter symbol R is used to indicate *resistance* in schematic diagrams.

When the ohm is too small a unit for practical use we employ either the *kilohm* (kΩ), which is equal to one thousand (1,000) ohms, or the *megohm* (MΩ), which is equal to one million (1,000,000) ohms.

All components used in electric circuits contain some resistance. Of particular interest to electricians is the resistance of the conductors.

Resistance of metallic conductors

Four factors affect the resistance of a metallic conductor: *length, cross-sectional area* (thickness), *the type of material from which the conductor is made,* and *temperature.*

The resistance of a metallic conductor is directly proportional to its length. That is, as the length of the conductor increases, its resistance increases. The truth of this statement is apparent when we consider that the farther the free electrons move through the conductor, the more collisions there are with ions; hence, the greater the resistance.

The resistance of a metallic conductor is inversely proportional to the cross-sectional area (thickness). That is, as the thickness of the conductor increases its resistance decreases, and as its thickness decreases, its resistance increases.

The resistance of a metallic conductor is dependent on the type of material used to form the conductor. The truth of this statement can be seen when we recall that the distinction between conductor and insulator materials is based on the number of free electrons that the material of the one or the other possesses per unit volume.

For most conductors, resistance increases when the temperature rises and decreases when the temperature drops. A good example of this is the incandescent lamp with a tungsten filament. The *cold* resistance of the filament is about 18 ohms, but its *white hot* resistance is about 240 ohms. As a result of its relatively low cold resistance, the tungsten filament experiences a heavy *surge* of current initially. As the filament heats and its resistance increases, current gradually declines to a normal value. In some devices, such as the heating elements of electric stoves, heavy surges cannot be tolerated and special metal alloys, such as Nichrome II Ⓣ, whose resistance value remains reasonably constant over a wide temperature range, must be employed.

Measuring resistance

The instrument most often used for the direct measurement of resistance is called an *ohmmeter*. The ohmmeter contains its own source of EMF (electromotive force) in the form of a battery so it is never used to make measurements in a *hot* (energized) circuit. *All power to the circuit under test must be turned off before any attempt is made to connect the ohmmeter into the circuit.* Failure to do so will result in destruction of the meter.

An ohmmeter is connected *across* the component or circuit whose resistance is to be measured.

Since ohmmeters with different scale arrangements and different control arrangements are available commercially, no attempt at physical description will be made here. Again, be sure to use the manufacturer's literature supplied with the instrument as a guide to its use.

Multimeters

To avoid carrying around separate ammeters, voltmeters, and an ohmmeter, most electricians use a meter called the *multimeter* which incorporates all three meters into a single instrument. The quantity to be measured is selected by means of a *function switch*. A *range switch* provides maximum flexibility.

1.8   *Ohm's Law*

In 1825 a German scientist, Georg Simon Ohm, performed a series of experiments that led to a statement of one of the most important laws that govern the behavior of all electric circuits. Working with a circuit similar to that shown in Fig. 1-9, Ohm discovered that every time he closed the switch, the circuit current had the same constant value. He also found that, *as long as the temperature of the conductors did not change,* the circuit current was *directly proportional* to the applied EMF. In other words, if the EMF was *doubled,* the current was also *doubled.* Conversely, if the EMF was reduced to *one-half* its original value, the current also fell to *one-half* its original value.

Carrying his experiment one step further, Ohm discovered that the value of current changed whenever he changed either the *type* or *thickness* or *length* of the conductor, or the *type* of bulb. Why did these changes occur? He found the answer, of course—that every time the conductor or the bulb was

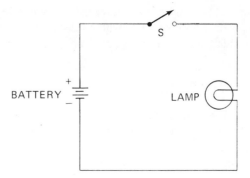

Figure 1-9   Circuit similar to that used by Ohm

changed, the resistance of the circuit changed. When the resistance of the circuit *increased,* the current *decreased.* When the resistance of the circuit *decreased,* the current *increased.*

On the basis of these discoveries, two very definite statements can be made about the behavior of electric circuits. (1) When the resistance and temperature of the circuit are constant, current is directly proportional to voltage. If the voltage increases, current increases and if the voltage decreases, current decreases. (2) When the source of EMF provides a constant potential difference in the circuit, and the temperature is constant, current is inversely proportional to resistance; when the resistance goes up, the current goes down and when the resistance goes down, current goes up.

It was to honor the memory of George Ohm that the unit of resistance was called the *ohm,* and the relationship between voltage, current, and resistance is summed up in *Ohm's law.* The three forms of Ohm's law follow:

$$R = \frac{E}{I}; \qquad\qquad I = \frac{E}{R}; \qquad\qquad E = IR \qquad\qquad (1\text{-}1)$$

where R is the resistance in *ohms,* E is the applied EMF in *volts,* and I is the current in *amperes.*

A few simple examples will serve to demonstrate the usefulness of Ohm's law. In each case, the circuit will be similar to that shown earlier in Fig. 1-9.

Let $E = 30$ V and $I = 3$ A. What is the value of R?

$$R = \frac{E}{I} = \frac{30}{3} = 10 \ \Omega$$

Let $R = 20 \ \Omega$ and $E = 40$ V. What is the value of I?

$$I = \frac{E}{R} = \frac{40}{20} = 2 \ A$$

Let I = 3 A and R = 20 Ω. What is the value of E?

$$E = IR = 3 \times 20 = 60 \text{ V}$$

1.9   Power

In everyday life, it is important to know *how much work* is accomplished by a given process and *how long it takes* to do a given amount of work. The same situation holds true in dealing with electric circuits.

In electric circuits, *the rate of doing work* is called *power* and is indicated by the letter symbol P.

To honor the memory of James Watt, inventor of the steam engine, the *unit* of power was named the *watt,* abbreviated **W**.

To calculate power in an electric circuit we use the relationship

$$P = EI \tag{1-2}$$

where P is the power in watts, E is the applied EMF in volts, and I is the circuit current in amperes. If several load devices are connected in the circuit, and you wish to determine the power dissipated by a single device, V (voltage drop) is used in place of E in Equation 1-2.

Suppose a lamp is connected to a 120 V source and draws a current of 0.5 **A**. To determine the rate at which electric energy is supplied to the lamp we use Equation 1-2.

$$P = EI = 120 \times 0.5 = 60 \text{ W}$$

Electric lamps, motors, and clothes dryers are just a few of the devices an electrician encounters that are rated in watts. The *wattage rating* of any device *indicates the rate at which the device converts energy from one form to another* and often indicates the *safe operating limit* of the device.

Since power is dissipated by the resistance of any practical electric circuit, it is often convenient to express power in terms of R. By Ohm's law, $E = IR$. If we substitute IR for E in Equation 1-2, it becomes

$$P = EI = IRI = I^2 R \tag{1-3}$$

If, for example, the resistance of the lamp filament in the previous example is 240 Ω, the power dissipated by the lamp is

$$P = I^2 R = 0.5^2 \times 240 = 60 \text{ W}$$

Since P = 60 W in both examples, it is apparent that Equations 1-2 and 1-3 can be used interchangeably.

The above example shows that power developed in a circuit of *fixed* resistance is *directly proportional* to the *square* of the *current*. If the current is doubled, the power is increased by a factor of 4, since $2^2 = 4$. If the current is tripled, the power is increased by a factor of 9, since $3^2 = 9$. By similar reasoning, if the current is halved, the power is reduced to one-fourth of its original value, and so on.

We can derive another useful relationship from Equation 1-2 by substituting E/R for I. This gives

$$P = EI = E \times \frac{E}{R} = \frac{E^2}{R} \qquad\qquad (1\text{-}4)$$

Now, if a 240 Ω lamp is connected to a 120 V source of EMF, the rate at which the lamp uses power is

$$P = \frac{E^2}{R} = \frac{120^2}{240} = 60 \text{ W}$$

The above example shows that Equations 1-2, 1-3, and 1-4 may be used interchangeably. Which equation you choose is determined by the known quantities. In practice, the value of any one quantity may be determined by measurement or, in most cases, by reference to the stamp or rating plate found on most electrical devices.

The last example also shows that the power developed in a circuit of *fixed* resistance is *directly proportional* to the *square* of the *applied voltage*. Thus, when the voltage is doubled, power increases by a factor of four. Similarly, when the voltage is halved, power decreases to one-fourth its original value.

In working with practical circuits, always keep in mind that power is directly proportional to the square of *either voltage* or *current*. Thus, if a circuit is modified in any way and the value of circuit resistance is changed, *before* power is applied to the circuit, make sure the wattage ratings of all devices connected thereto will not be exceeded.

Because the watt is a very small unit, it is often more convenient to express power in multiples of one thousand watts (*kilowatts*) or kW.

Measuring power

Although we can calculate the power in a load by taking separate ammeter and voltmeter readings, a direct-reading *wattmeter* is a very useful instrument, particularly for ac circuits.

A wattmeter containing an *electrodynamometer movement* can be used in either ac or dc circuits. Inside the instrument case there are two *coils*—a *current* coil and a *voltage* coil. To assist the user in connecting a wattmeter properly, it is customary in practice to identify one end of each coil with ± marks. Standard connections require that we connect the identified terminal of the current coil to the generator and the identified terminal of the voltage coil to the side of the line containing the current coil.

More will be explained about the use of wattmeters in Chapter 2. Before you attempt to use any wattmeter, be sure you read the manufacturer's literature supplied with the instrument.

1.10   *Horsepower*

Electric motors, except for a few miniature types, are rated in *horsepower,* abbreviated hp. When it is necessary to convert from horsepower to watts, or vice versa, use the relationship

$$1 \text{ hp} = 746 \text{ W}$$

Theoretically, then, the rate at which electric energy is supplied to a 1 hp motor is 746 W. In actual practice, however, the motor will consume more than 746 W from the source. The difference between the theoretical and actual power requirements is consumed in overcoming such things as bearing friction and the air resistance of moving parts. Typical motor consumptions in watts are shown in Table 1-2.

Motor size in hp	Typical wattage consumption
1/4	300 to 400
1/2	450 to 600
Over 1/2, per hp	950 to 1100

Table 1-2   Typical power consumption of motors

1.11   *The Watthour*

To measure the theoretical electrical energy consumed by all devices connected to an electric circuit involves the determination of *how much power* is used and for what *length of time.* The *watthour,* abbreviated Wh, may be used for this purpose and is simply the product of power in watts and time in hours. More often, however, use is made of the *kilowatthour,* abbreviated kWh, where kilowatts are used in place of watts.

Suppose, for example, a 250 W lamp burns for 10 hours. The power consumed by the lamp is

$$250 \times 10 = 2500 \text{ Wh}$$

or, since 250 W = 0.25 kW,

$$0.25 \times 10 = 2.5 \text{ kWh}$$

It is on the basis of the kWh that we pay our electric bill. To illustrate, suppose six lamps, each rated at 100 W, operate 8 hours per day for 30 days, and that the utility company charges $0.02 per kWh. The cost to operate the lamps is determined as follows:

$$\text{Total wattage} = 6 \times 100 = 600 \text{ W} = 0.6 \text{ kW}$$

$$\text{Daily kWh} = 8 \times 0.6 = 4.8 \text{ kWh}$$

$$\text{kWh for 30 days} = 30 \times 4.8 = 144 \text{ kWh}$$

$$\text{Cost} = \$0.02 \times 144 = \$2.88$$

A kWh *meter* is installed in all residences and is read monthly by a representative of the local electric supplier.

Some kWh meters have a *cyclometer* dial similar to that shown in Fig. 1-10(a), where the indicated number of kWh is 2436. If the same meter is read one month later and shows 2981, for example, the electric consumer is charged for 545 kWh or the difference between 2981 and 2436.

Other kWh meters contain four dials which are marked as shown in Fig. 1-10(b). From left to right, the *first* and *third dials* are read in a *counterclockwise* direction while the *second* and *fourth dials* are read in the *clockwise* direction. The meter reading is determined by the *last* number the pointer has *passed* on each dial. On the first dial, the last number passed is 4. On the second dial the pointer rests directly on 4; however, until the pointer on the third dial passes zero the proper reading for the second dial is 3. The third dial reads 8 and the fourth dial 7. Thus the total indicated kWh is 4387.

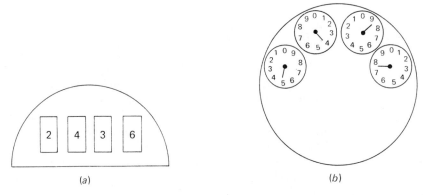

(a)

(b)

Figure 1-10   kWh meters with (a) cyclometer dial and (b) four dials

1.12   *Series Circuits*

An electric circuit in which all the component parts are connected *end-to-end,* as in Fig. 1-11, is called a *series* circuit.

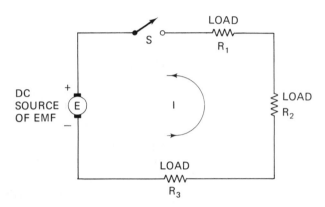

Figure 1-11   A series circuit

To work intelligently, an electrician must be able to make certain tests in any type of circuit. On the basis of these tests he must then decide if the circuit is operating properly or if a *fault* exists at some point in the circuit. If a fault exists, he must be able to find and correct the fault quickly.

Fortunately, the distribution of voltage, current, and resistance in electric circuits is not random but follows definite patterns.

Distribution of *E, I,* and *R*

Let us first see how *E, I,* and *R* are distributed in a properly operating series circuit. Later we will consider what happens when this distribution is upset by a circuit fault.

1.   In a complete series circuit, that is, one in which all switches and other control devices are closed, current has but *one* path to follow: from the negative terminal of the source, through each load device in turn, and back to the positive terminal of the source. Thus, *the same current exists in all parts of a series circuit;* accordingly, an ammeter connected to any point in the circuit will give the *same* reading.

2.   The component parts of a series circuit are connected end-to-end and the same current passes through each part in turn. Since, with a given source voltage, the circuit current is determined solely by the resistance, it should be apparent that *the total resistance of the circuit is equal to the sum of the individual resistances.*

Suppose, for example, three load devices are connected in series with a source of EMF and the resistances of these devices are 75 Ω, 200 Ω, and 165 Ω respectively. Then, the total resistance of the circuit is 75 + 200 + 165 = 440 Ω.

3.   In a series circuit, *the sum of the voltage drops, across the individual load resistances must equal the applied EMF.*

If, for example, 75 Ω, 200 Ω, and 165 Ω load devices are connected in series with a 220 V source of EMF, the circuit current, by Ohm's law, will be $I = E/R = 220/440 = 0.5$ A. Then, since the same current passes through each load device in turn, the voltage drop across the devices ($V = IR$) will be 37.5 V, 100 V, and 82.5 V respectively. The sum of these voltage drops is 220 V, the value of the applied EMF.

4.   If the value of any resistance is changed, while the applied EMF is held constant, the values of the total resistance and the circuit current must also change. Thus, *the change in resistance value of a single component in a series circuit results in a complete redistribution of voltage in the circuit.*

To illustrate we will again use the 75 Ω, 200 Ω, and 165 Ω load devices. Let us suppose, however, that for some unknown reason the resistance of the 200 Ω device changes to 100 Ω. Then, the total resistance in the circuit will be 75 + 100 + 165 = 340 Ω, and the circuit current will be 220/340 = 0.647 A. The voltage drops across the individual devices becomes 48.5, 65.0, and 106.5 V, respectively. Notice that the sum of the drops is still equal to the applied EMF even though a redistribution of voltage has taken place in the circuit.

Series-circuit faults

Now let us look at common circuit faults and their effect on circuit operation.

incorrect EMF—If the applied EMF is too *low,* the circuit current will also be abnormally low. Since $P = I^2R$, load devices will not develop the power required for proper operation. If the applied EMF is too *high,* the circuit current will be abnormally high. Thus, the power dissipation capability of one or more devices connected in the circuit may be exceeded, causing the device to *burn out.*

wrong replacement part—If the resistance of a replacement part is lower than that of the original component, circuit current will increase. Since $P = I^2R$, one or more of the devices connected in the circuit may be burned out.

If the resistance of a replacement part is considerably higher than that of the original component, current will decrease. Again, since $P = I^2R$, the devices connected to the circuit may not be able to develop the power required for proper operation.

Finally, even if the resistance of the replacement part is correct, its power dissipation capability must be *equal to* or *greater than* that of the original part.

To provide a *safety margin,* the wattage rating of any device should be higher than its maximum anticipated dissipation. To be on the safe side, always use exact replacement parts if the circuit operated properly before a fault occurred.

open circuit—An *open circuit,* usually referred to as an *open,* is simply a break in the conducting path. The resistance of an open circuit is infinite for all practical purposes, since air replaces the conducting material. Air, you will recall, is an excellent insulator.

When an open occurs in a series circuit, there is no complete path through which current can pass. The current in all parts of the circuit is therefore zero. For example, if the circuit contains a string of series-connected lamps and the filament of one lamp burns out, all the lights are extinguished since the circuit current is zero.

short circuit—A *short circuit,* usually referred to as a *short,* occurs when a portion of the resistance normally encountered in an electric circuit is bypassed or short-circuited.

Consider what happens when two bare conductors come into contact with one another at points *A* and *B*, thereby forming a short. See Fig. 1-12. The only resistance between points *A* and *B*, when this occurs, is the very small resistance of the wire. As a result, essentially the total *line* (circuit) current passes from point *A* to point *B*; the current through resistances R_2, R_3, and R_4

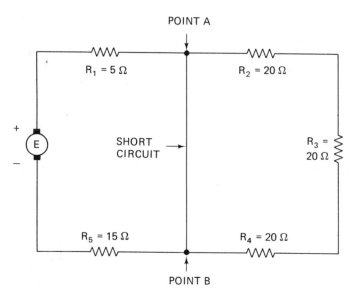

Figure 1-12   A short circuit

drops to zero—for all practical purposes. (Notice the use of numerical subscripts to distinguish one resistive load device from another.) The effective removal of R_2, R_3, and R_4 from the circuit causes a sharp increase in current. To illustrate: the normal circuit current is 1.5 A (E divided by the total resistance). When the short occurs, the total resistance drops to 20 Ω ($R_1 + R_5$) and the current becomes 6 A.

Before the short occurs, the power dissipated by R_1 *and* R_5 is 11.25 and 33.75 W, respectively ($P = I^2R$). Under short-circuit conditions, R_1 dissipates 180 W and R_5 dissipates 540 W. Unless R_1 and R_5 are capable of handling such excessive amounts of power—which is extremely unlikely—these elements will be destroyed by the excessive heat. Once the devices represented by resistance R_1 and R_5 are burned out, an open circuit exists.

To summarize: in a *series* circuit, a short usually results in the destruction of those devices connected between the source of EMF and the point at which the short occurs. It should be understood, of course, that the source of EMF itself may also be either damaged or destroyed—depending on the nature of the source.

In view of the dangerous conditions that can result from a short circuit, extensive use is made of *protective devices* that open the main line automatically and disconnect the source of EMF from the circuit in which the short occurs. We will examine these devices in Chapter 5.

1.13   *Parallel Circuits*

A *parallel* circuit, such as that shown in Fig. 1-13, is one in which the components are connected in a *side-by-side* manner and where *more* than one path exists for the passage of current. Each current pathway is referred to as a *branch*.

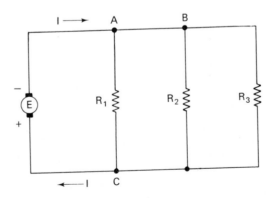

Figure 1-13   A parallel circuit

Distribution of *E, I,* and *R*.

Let us first see how *E, I,* and *R* are distributed in a properly operating parallel circuit and then what happens when this distribution is upset by a circuit fault.

1.   Referring again to Fig. 1-13, we notice that each of the three resistive load devices, R_1, R_2, and R_3, is connected directly across the terminals of the source of EMF. Thus, *the same voltage* (that of the source of EMF) *exists across all branches of a parallel circuit.*

2.   Also, notice in Fig. 1-13 that a *separate* current passes through each branch of the parallel circuit. The value of each individual branch current is calculated from Ohm's law by using the familiar relationship $I = E/R,$ where *E* is the source of EMF and *R* is the resistance of the branch.

3.   Finally, notice that the total current must pass from the negative terminal of the source of EMF to point *A.* At point *A,* the current divides with a portion passing through branch R_1 and the remainder going to point *B.* At point *B,* the current again divides with a portion passing through branch R_2 and the remainder through branch R_3. At point *C,* the currents passing through all three branches recombine and return to the positive terminal of the source of EMF. From this we conclude that *the total current in a parallel circuit is equal to the sum of the individual branch currents.*

4.   In a *series* circuit, the equivalent resistance (that is, the total effective resistance) was found to be equal to the sum of the individual resistances. In a *parallel* circuit, a much different condition holds true. *Whenever two or more branches are connected in parallel, the total equivalent resistance of the circuit is always less than the value of resistance in the branch having the least resistance.*

EXAMPLE: If three branches are connected in parallel across a given source of EMF and the resistances of the branches are, say 25, 50, and 75 ohms, the total effective resistance of the circuit, as far as the source is concerned, is *less* than 25 ohms.

With a little thought, this seemingly strange state of affairs comes as no great surprise. Because the full applied EMF is across each branch, more current is required from the source, and the equivalent resistance of the circuit must decrease to meet this demand for increased current.

Since the practical electrician is seldom, if ever, called upon to compute the total effective resistance of a parallel circuit, we will not spend time in studying the equations needed for such computations. The effect of adding parallel branches should, however, be kept in mind when working with circuits of this type.

Parallel circuit faults

In a parallel circuit, the effects of an incorrect EMF or a wrong replacement part are essentially the same as in a series circuit.

open circuit—In a parallel circuit, the *location* of the open determines what effect it has on circuit operation. In Fig. 1-14, where six lamps are connected in three parallel branches, an open in the main line at point *A* has the same effect as in a series circuit, because the total current passes through point *A*. Thus, an open at point *A* would cause *all* of the lights to be extinguished. On the other hand, if the filament of any *one* lamp burned out and created an opening in *one* of the branches, both of the lamps located in that branch would be extinguished because the current in that branch would be zero. All of the remaining lights would continue to glow, however, because each of the branches in which they were located would still be connected across the source of EMF and would continue to form a complete path for the passage of current. *An opening in any one branch has no effect on any other branch.* The current in each remaining branch stays the same because $I = E/R$ and neither E nor R changes. The total circuit current decreases, however, by an amount equal to the current that normally passes through the branch in which the open occurs.

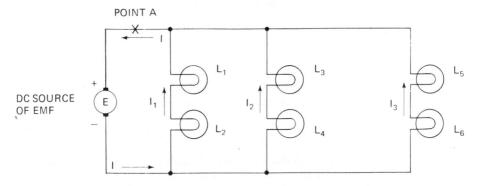

Figure 1-14   In a parallel circuit, the effect of an open depends upon its location

short circuit—In the parallel circuit of Fig. 1-15, a short is assumed to exist between points *A* and *B*. Because the resistance of the short-circuit path is practically zero, essentially the entire circuit current passes through the source of EMF, through the conductors connecting points *A* and *B* to the source, and through the short circuit itself. Because the effective circuit resistance is practically zero, the current reaches a very high value of, perhaps, several hundred amperes; and it produces so much heat that the source, the conductors, or both, may ignite and burn.

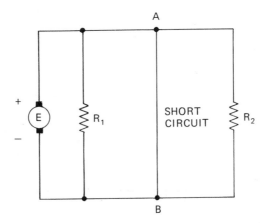

Figure 1-15   A short-circuited parallel circuit

The devices represented by resistances R_1 and R_2 in Fig. 1-15 are made inoperative as a result of the short circuit since the current through these branches drops to essentially zero; however, the devices are not damaged and may be reused once the circuit fault is repaired.

To sum up, a short in a parallel circuit causes a dangerously high-line current that may either damage or destroy the source. In any event, rapid and severe overheating of the source and conductors between the source and the short will occur—and a fire is very possible. The devices connected in the normal branches are not affected by the short, but they are made inoperative until the circuit fault is corrected.

1.14   *Series-Parallel Circuits*

So-called *series-parallel* circuits are simply a combination of the series and parallel arrangements. A typical series-parallel circuit is shown in Fig. 1-16. Since the total circuit current passes through R_1, the device represented by this resistance is in *series* with the source of EMF. At the junction of R_2 and R_5, the total current splits, with part of it passing through branch *A* and the remainder through branch *B*. Notice, however, that branch *A* contains three series-connected resistances while branch *B* contains four series-connected resistances. When the two branch currents recombine they flow through R_9. Thus, R_9 is a *series*-connected component.

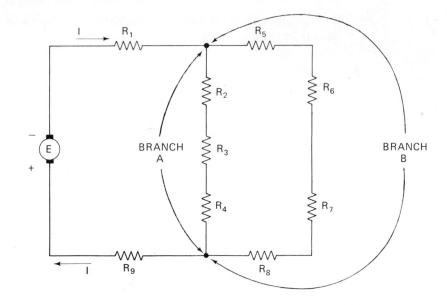

Figure 1-16   Series-parallel circuit

The *same* voltage is impressed across both the parallel branches. The value of this voltage is E minus the voltage drop occurring across resistance R_1. If, for example, V_1 is 20 V and E is 120 V, the voltage appearing across both branches is $120 - 20 = 100$ V.

Although, series-parallel circuits may appear to be somewhat involved, just remember that all the characteristics of series circuits apply for all series-connected components. Similarly, all the characteristics of parallel circuits apply when the components are in parallel.

1.15   *Power in Series and Parallel Circuits*

Whenever current passes through a resistance, a certain amount of power is dissipated. This power must come from the source; the circuit arrangement cannot change this requirement in any way. Thus, regardless of whether resistances are connected in series, parallel, or some combination thereof, the total power taken from the source is always equal to the sum of the individual powers dissipated by the circuit components. Since the total power requirement is supplied by the source by means of its total line current, the total power requirement is always equal to the product of E and I (total).

1.16   *Magnetism*

The ancients knew that special properties characterized *lodestones* which were magnetic iron ores. Two of those magnetic forces which are utilized by modern technology follow: (1) When a piece of lodestone is suspended so that it can swing freely in a horizontal plane, it invariably assumes a definite position relative to the earth's North and South directions; and (2) lodestone specimens attract small fragments of iron.

The attractive force which a bar magnet exerts on iron filings is concentrated near its ends. These ends are called the *magnetic poles* of the bar magnet. When the bar is pivoted horizontally, the end pointing approximately toward the geographic North of the earth is called the north-seeking pole or simply the *north pole* of the magnet and the other, south-seeking end, is the *south pole* of the magnet. Any fragment of a magnetic substance, no matter how small, has both a north and a south pole. A single magnetic pole *cannot* exist.

If two compass needles are brought close to each other, the following law of magnetic poles is demonstrated: *like poles repel; unlike poles attract.*

Only a few materials exhibit marked magnetic properties. *Iron* and *steel* and their *alloys* are strongly magnetic; on the other hand, *cobalt* and *nickel* have very negligible magnetic properties in comparison with iron. *Soft* iron is more readily magnetized than steel, but steel retains its magnetism to a greater extent than soft iron. Consequently, magnets made of steel or steel alloys are good *permanent* magnets. Soft-iron magnets are good *temporary* magnets, but they either lose or reverse their polarity quickly. When a compass needle is brought near a magnet, a *force* acts on the needle and the needle assumes a definite direction with respect to the magnet.

The region about a magnet in which this force is present is called a *magnetic field* and is made up of *magnetic lines of force,* also called *flux.*

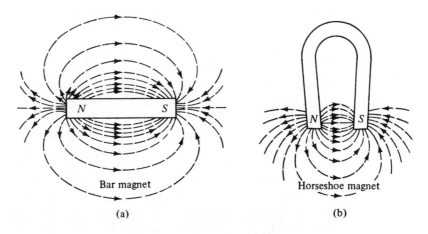

(a)   (b)

Figure 1-17   Magnetic fields of (a) bar and (b) horseshoe magnets

Since the compass needle assumes a particular position when placed near a magnet, a magnetic field has *direction.* The direction of a magnetic field at a certain point is defined as the direction indicated by the north pole of a compass needle when placed at this point.

It is convenient to represent the direction and intensity of a magnetic field by lines of force. The *intensity* of a field is represented by the density of the lines and the *direction* by arrow heads on the lines. See Fig. 1-17. All lines of force are *closed* paths directed from the north pole to the south pole in the region *outside* the magnet and returning *inside* the magnet from south to north.

The field existing between two like poles (which repel) is shown in Fig. 1-18.

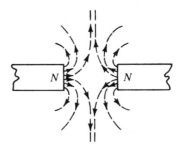

Figure 1-18   The field existing between two like poles

1.17   *Magnetic Fields Around Conductors*

If a *straight* conductor carrying a steady current is brought near a compass needle, the needle is deflected. There is no deflection if current does not flow. Therefore a magnetic field exists about the conductor only when a current flows through it. By moving the compass needle around the conductor and noting the direction in which the North Pole points, the lines of force of the field are observed to form concentric circuits about the conductor. See Fig. 1-19(a). If the direction of the current is reversed, the lines of force reverse too. See Fig. 1-19(b).

The following *left-hand* rule relates the direction of the lines of force to the direction of current. *If the conductor is grasped in the left hand with the thumb pointing in the direction of electron flow, the fingers point in the direction of the magnetic field around the conductor.*

The intensity of the field is greatest close to the conductor and rapidly decreases with distance from the conductor. The intensity of the field is increased if the current strength is increased.

The intensity of the magnetic field about a straight conductor is very

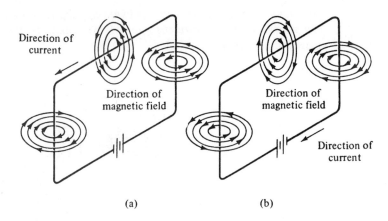

Figure 1-19   Magnetic field existing around a straight conductor carrying a steady current: (a) current moving right to left, (b) current moving left to right

small even with large currents. When, however, the conductor is bent in the form of a loop, as in Fig. 1-20, a much stronger field is produced inside the loop, since the lines of force established in each small section of the conductor combine as they flow in the same direction in the center of the loop.

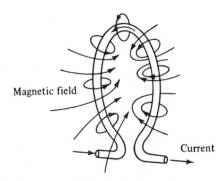

Figure 1-20   Magnetic field around a conductor bent in the form of a loop

If the conductor is coiled into a number of turns as in Fig. 1-21, the lines of force within each loop combine to produce a still larger field. Such a coil or helix with its length great as compared to its diameter is called a *solenoid*. The diagram of Fig. 1-21 shows that the field about a solenoid is like the field about a bar magnet. The ends of the coil from which the lines emerge acts as a north pole and the end into which they enter acts as a south pole. The following rule relates current direction in the solenoid to the polarity of its ends. *If the conductor is grasped in the left hand with the fingers pointing in the direction of electron flow, the outstretched thumb will point to the north pole of the magnetic field around the coil.*

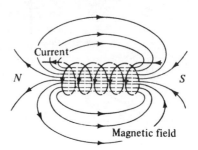

Figure 1-21   Magnetic field around a solenoid

If a solenoid is supplied with a core of magnetic material, its magnetic field is *strengthened*. The combination of a coil and its magnetic core is called an *electromagnet*. The fact that the electromagnet is a temporary magnet is a distinct advantage. That is, it is surrounded by a magnetic field only as long as current flows through its turns. That permits *on* or *off* control of the magnetic field merely by the flick of a switch.

If a solenoid is provided with a *movable* soft-iron core, called a *plunger,* when a current is made to flow through the turns of the coil, the magnetic field tends to *pull* the plunger into the center of the coil. The coil together with its movable core is usually called a *solenoid* (although, strictly speaking, only the coil is the solenoid). A solenoid can also be arranged so that its plunger is pushed *out* of the center of the coil when current through the coil is discontinued.

1.18   *Electromagnetic Induction*

Since a magnetic field is always associated with an electric current, it seems logical that a current is always associated with a magnetic field. The following experiments show that this is true, providing certain conditions are fulfilled.

A coil of wire is connected to a galvanometer which is a very sensitive current-measuring instrument. Then a magnet is moved into and out of the coil as indicated in Fig. 1-22. The following data can be observed: (1) as the magnet moves into and out of the coil, a current flows, as indicated by the galvanometer's recordings; (2) when the magnet is at rest there is no current; (3) the more rapidly the magnet is moved, the greater the strength of the current measured by the galvanometer; and (4) if an unmagnetized bar of iron is moved in and out of the coil there is no current.

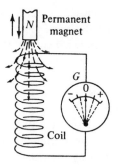

Permanent magnet

G

Coil

Figure 1-22   Electromagnetic induction

The current flow that is observed must be due to an EMF. This EMF is produced in the coil *only* when the *moving* magnetic field of the magnet is *cutting* across the turns of the coil. Furthermore the strength of this EMF depends on the *speed* of the moving field. An EMF that is set up in a conductor when it is cut by a moving field is called an induced EMF and the process is known as *electromagnetic induction.* If the magnetic field is held in a fixed position and the coil moved up and down over the magnet the same observations can be made.

The results of the experiments described above are summed up in the two laws of induction known as Faraday's laws. (1) *When the magnetic flux (lines of force) linked with a circuit changes, an EMF is induced in the circuit.* (2) *The strength of the induced EMF is proportional to the rate-of-change of flux linkage.*

In dealing with electromagnetic induction caused by a changing current rather than by mechanical motion, Faraday's second law may be restated as follows: *the magnitude (strength) of the induced EMF is directly proportional to the time rate-of-change of current.*

In Fig. 1-23 coil A is placed within coil B with no electrical connection between the two coils. A galvanometer is connected in series with coil B. When

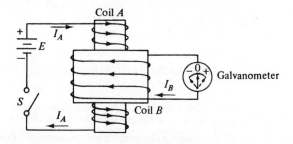

Coil A

I_A

E

I_B Galvanometer

S

I_A Coil B

Figure 1-23   Demonstrating Lenz's law

switch S is closed, current I_A flows through coil A in a counterclockwise direction. At the same time, the galvanometer indicates a current, I_B, through coil B. This induced current lasts only a very short time and dies out quickly, even though current I_A is maintained steady by the battery.

When switch S is opened, the galvanometer again shows an induced current, but now the direction of the current is reversed; that is, I_B now flows in the same direction as I_A. As before, current I_B dies out rapidly.

Notice that the induced current, I_B, exists when, and only when, there is a change in current I_A. This phenomenon can be further verified experimentally by inserting an appropriate ammeter in series with coil A so that it can record the fact that I_A builds up gradually to a final steady value, and that I_B flows only during this buildup of I_A. Similarly, when switch S is opened, I_A does not die out immediately, but gradually, and I_B flows only during this time.

Also note that when switch S is closed, the direction of I_B is *opposite* to the direction of I_A. When the switch is open, the direction of I_B is the *same* as that of I_A.

In both cases the induced currents *oppose the change* that takes place in I_A. First, current I_B opposes the setting up of I_A, and then its opposes the stopping of I_A. As soon as the change in I_A is over, I_B dies out.

The behavior of the induced current can be best understood in terms of the magnetic fields set up by the currents flowing in the coils. A little reflection shows that when I_A begins, it sets up a north magnetic pole at the top of coil A. Current I_B opposes this action by setting up a south magnetic pole at the top of coil B, tending to neutralize the field set up by I_B. Once the field is set up by I_A, current I_B ceases to flow. When switch S is opened, the field set up by I_A begins to decay, and I_B once again produces a field that opposes this decay. In each case, I_B sets up a field that opposes the change taking place within the cores of the two coils as a result of the change in current I_A.

This phenomenon has come to be known as Lenz's law, which may be generalized as follows: *An induced current always flows in such a direction that the magnetic field it produces opposes any change in the existing field produced by the inducing current.*

The important concept in this generalization is that of *change*. The field of the induced current *always* opposes any change taking place in an existing field.

1.19   *Mutual Induction and Mutual Inductance*

In Fig. 1-24, two coils are placed near each other. The *primary* coil is connected in series with a dc source of EMF and a *rheostat*. In electric circuits, resistance is used to control (limit) the amount of current. Although we could use high-resistance conductors to place the necessary resistance in the circuit, this would not be a very practical solution. Instead, the extra resistance is lumped into a single component called a *resistor*. Remember that resistance is a *property* of a circuit while a resistor is a *component* used in a circuit. Since resistance is used to control current, heat is generated when current passes through a resistor; therefore, resistors are made in various sizes and shapes so that they can safely dissipate a certain amount of heat. The correct size must be selected for a given application.

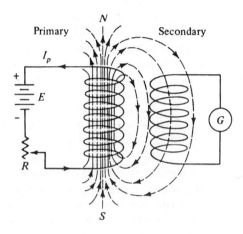

Figure 1-24   Mutual induction

A rheostat is simply a special form of resistor. One connection is made to an end of the rheostat while the other connection is made to a movable contact located inside the rheostat casing. As the arm (wiper) moves across the resistance material, the value of the resistance is varied. Thus, a rheostat is simply a *variable* resistor.

If connections are made to both ends and to the wiper of a variable resistor, the device is called a *potentiometer.*

Returning to Fig. 1-24, the *secondary* coil has a galvanometer connected across its terminals. If the primary current is held constant, then the magnetic field about the primary coil is constant. The magnetic lines about the primary coil link the secondary coil. Since the secondary coil is not being cut by a moving field; however, there is no EMF induced and there is no secondary current.

Now, if the primary current is varied by means of the rheostat, the field of the primary changes. As the primary current increases, the field increases; the magnetic lines spread outward across the secondary coil and an EMF is induced in the secondary circuit. As the primary current decreases, the field decreases; the magnetic lines collapse inward across the secondary coil and an EMF is induced in the secondary circuit. If the rate of change of the primary current is made greater, more lines cut across the secondary coil per second and the induced EMF is greater. These observations confirm Faraday's first and second laws.

The process by which an EMF is induced in a secondary winding by a changing current in a primary winding is called mutual induction, and the coils are said to possess mutual inductance.

Before we discuss the factors which determine the mutual inductance of two coils, let us first define a term called *permeability*. By definition, *permeability is a figure that indicates the ability of a material to permit the setting-up of magnetic lines of force.*

The permeability of a *vacuum*—and, for practical purposes, air and all other nonmagnetic substances—is rated as 1. Thus, permeability is the ratio between the magnetic flux produced in a magnetic material and the flux that would be produced if air were used instead, the so-called magnetizing force remaining constant.

Not all magnetic substances have the same permeability. For example, the permeability of soft iron, is much greater than that of steel, and there are some magnetic alloys that have a much greater permeability than soft iron.

Now, the mutual inductance of two coils depends on the following interacting factors: (1) *construction*—mutual inductance increases with each increase in the number of turns per unit length (by the inch, for example) in each coil and also with any increase in the cross-sectional area or thickness of each coil; (2) *permeability*—an increase in the permeability of the cores or of the medium about the coils increases mutual inductance; and (3) *relative position* of the two coils—the greater the distance between the coils, the less the mutual inductance; conversely, lesser distance means improved mutual inductance. Also, the mutual inductance is greatest when the two coils are coaxial and least when their axes are at right angles to each other.

1.20   Self-Induction and Self-Inductance

When a changing current passes through a conductor, a magnetic field grows and collapses through and about the conductor. Since the conductor is cut by its own changing field, we know from Faraday's law that an EMF must be induced in the conductor. This process is called *self-induction.*

Now, according to the further developments noted in Lenz's law, the induced EMF must act in a direction that opposes an applied EMF. Hence, the induced EMF is called a *counter EMF* abbraviated CEMF.

When the CEMF opposes the applied EMF, it must also oppose any change in current. *The property of an electric circuit that opposes any change in current in that circuit is called self-inductance or, simply, inductance.*

It must be realized that any conductor, whether in the form of a straight wire or a coil, possesses inductance. It must also be realized that the opposition offered to any change in current by inductance is a temporary condition. Thus, inductance must never be confused with resistance. When the current eventually reaches a steady value, the lines of force around a conductor are no longer expanding or collapsing and a CEMF is no longer present. Thus inductance may be thought of as a *temporary* opposition to the establishment of the maximum value of current.

The inductance of a coil is increased substantially by increasing the number of turns on the coil and also by increasing the permeability of the core.

1.21   *Precautions in DC Inductive Circuits*

In the circuit diagrammed in Fig. 1-25(a), a dc source of EMF, a switch, an iron-core coil (notice the symbol), and a resistance are connected in series. The resistance is that of the coil itself. When the switch is closed,

$$I = \frac{E}{R} = \frac{10}{2} = 5 \text{ A}$$

Next we see in Fig. 1-25(b) that an open switch represents a very high value of resistance. We call this resistance *leakage resistance*. Let us assume that the leakage resistance in this particular case is 0.4 MΩ. When the switch is opened, the 5 A current that is still in the coil must pass through the leakage resistance of the switch. The voltage drop across the leakage resistance and, therefore, across the switch contacts is

$$V = IR = 5 \times 400,000 = 2,000,000 \text{ V}$$

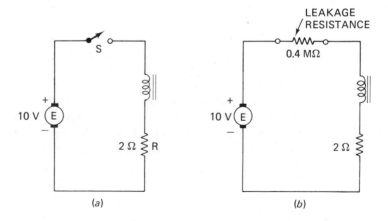

Figure 1-25   A dc inductive circuit: (a) switch open, and (b) leakage resistance of open switch

This *extremely high* voltage, which is equal to many times the value of the EMF that is applied when the switch is closed, *is matched by the CEMF generated in the inductor by the collapsing magnetic field.* Obviously, voltages of this magnitude are *dangerous.* Furthermore, voltages of this magnitude will cause a *breakdown* in the insulation of the coil and a severe arcing across the switch.

To avoid the creation of such high voltages when switch contacts are opened in dc circuits that contain a coil, a *discharge device* (such as the one called the *Thyrite* resistor) is connected across the coil. See Fig. 1-26.

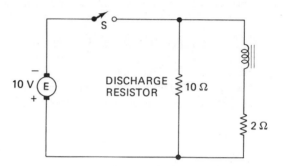

Figure 1-26   Use of a discharge resistor in a dc inductive circuit

When the switch in Fig. 1-26 is *closed*, a current of 1 A passes through the 10 Ω discharge resistor and the steady current through the branch containing the iron-core coil is 5 A.

When the switch is *opened*, the 5 A current through the coil must pass through the discharge resistor and create a voltage drop of 60 V. (The total resistance of the circuit with the switch open is 2 Ω + 10 Ω = 12 Ω, and I = 5 A.)

The direction of current produced by the CEMF is the same as the original current through the coil. Thus, the upper end of the discharge resistor becomes positive *with respect to* its lower end, and the voltage drop across the discharge path and the 10 V source of EMF are in *series* across the switch contacts. The total voltage drop across the switch contacts is 60 + 10 = 70 V. This is quite a reduction from the voltage appearing across the switch contacts when a discharge resistor was not used. The insulation of the coil is no longer subjected to abnormally high voltages, and there is no arcing across the open switch.

EXERCISES

1-1/   Describe a typical application in which energy is converted from one form to another.

1-2/   What happens when two *like* charges are brought close to each other?

1-3/   What happens when two *unlike* charges are brought close to each other?

1-4/   What do we mean when we say *polarity*?

1-5/   What is the *net* electrical charge of a normal atom?

1-6/   What is a *free* electron?

1-7/   What is a *conductor*? An *insulator*?

1-8/   Name the essential parts of any practical electric circuit.

1-9/   What is *current*? *Rate of flow of free electron.*

1-10/   Why is a source of electricity sometimes referred to as a *potential rise*? *E*

1-11/   What is the difference between the letter symbols *E* and V? *E = Rise   V = VD*

1-12/   How is a current-measuring instrument connected into a circuit? *DC → Polarity Req. (AC = Any)*

1-13/   How is a voltmeter connected into a circuit? *Source (+ & −) Across Loads +− VD*

1-14/   Define *resistance*. *— that which opposes the flow of current*

1-15/   What are the three basic forms of Ohm's law? *I = E/R*

1-16/   What factors determine the *resistance* of a metallic conductor? *AWG, Length, Temp, Type*

1-17/   Can a resistor of a given physical size be used in any electric circuit where its resistance value is correct? Explain your answer. *No, Depend whether Series, Parallel*

1-18/   Define electric *power*. *— E.M.F. (746 = 1 Hp) Energy consumed*

1-19/   Give *three* equations for computing power. *P = EI , P = I²R , P = E²/R*

1-20/   How is an *ohmmeter* connected into a circuit? *Power shut off and ...*

1-21/   What is a *series* circuit? A *parallel* circuit?

1-22/   What are the important characteristics of a *series* circuit?

1-23/   What are the important characteristics of a *parallel* circuit?

1-24/   How is a series circuit affected by an *open* circuit? A *short* circuit? *Blows Fuse or Tight*

1-25/   Define the terms north pole and south pole as applied to a magnet. State the law of magnetic poles and describe an experiment to demonstrate it. *Like Repel, unlike Attract.*

1-26/   Describe the magnetic field about a horseshoe magnet. *Flux Lines*

1-27/   Describe the magnetic field between two like poles. *Mutual Inductance*

1-28/   Describe the magnetic field around a straight conductor. *Self Inductance*

1-29/   What effect is produced when current is passed through a solenoid? *Inductance*

1-30/   The current through a coil is in a counterclockwise direction, as viewed from one end. What is the polarity of that end, and how is it determined? *Neg —*

1-31/   What factors determine the intensity of the magnetic field surrounding a coil?

1-32/   What is *permeability*?

1-33/   What is an *electromagnet*? What are its advantages compared to a permanent magnet?

1-34/   State Faraday's laws and explain their significance briefly.

1-35/   State Lenz's law and explain its significance briefly.

1-36/   Define *mutual* inductance.

1-37/   Define *self*-inductance.

1-38/   Distinguish between *mutual* induction and *self*-induction.

1-39/   Upon what factors does mutual inductance depend? Self-inductance?

1-40/   What precautions should be taken in a dc circuit containing one or more coils?

2

Alternating Current

2.1   *Introduction*

Recall from Sec. 1.6 of Chapter 1 that a current passing through a metallic conductor is the movement of free electrons under the influence of an applied EMF. Regardless of the polarity of the source of EMF, those free electrons *always* flow from negative to positive. Thus, if the source maintains a *fixed* polarity, the current always moves in the same direction and is called *direct current* (dc). If the source *reverses* its polarity periodically, however, the current also reverses direction at the corresponding periods because it must maintain the negative to positive convention. A current of this type is called *alternating current* (ac).

For a variety of reasons, the power-distribution systems of most—but not all—electric utility companies in the United States use alternating current. Let us take just a few moments to see why.

In the transmission of electric power from the generating station to the consumer, some waste (in the form of heat) occurs since even large-diameter conductors offer resistance to the passage of current. Remember that $P = I^2 R$. To reduce the power loss, the utility companies can either lower the resistance of the transmission lines by using thicker conductors or they can reduce the current passing through the conductors. If the resistance is lowered by half, the power loss is halved, but if the current is halved, the power loss is reduced by a factor of 4. Thus, a current reduction is the best way to reduce transmission-line losses. At the consumer end, however, large amounts of current are needed. Fortunately, both the low transmission-line current and high consumer-end current can be realized by using a device known as a *transformer*. Transformers are examined in detail later in Sec. 2.5. For the moment, let it suffice that you understand that these devices permit either the *stepping-up* or *stepping-down* of ac voltages and currents. Since the amount of power involved remains fairly constant—assuming minimum losses—and since $P = EI$, when the voltage is stepped-up, the current is stepped-down, and vice versa.

In a typical distribution system, power is generated initially at 13,800 V. By means of transformers, this voltage is then stepped-up to anywhere between 23,000 to 765,000 volts, depending on various factors, including the amount of power involved and the distance over which it must be transmitted. As noted earlier, when voltage is stepped-up, current is stepped-down. Thus current is reduced in the transmission line and losses are minimized by the systematic use of transformers at various points throughout the transmission system.

At an area substation, another transformer is used to step-down the extremely high transmission voltage to some lower value, such as 13,800 V. With the voltage back to its initial value, so also is the current—at least, for our purposes.

At the same substation additional transformers are often used for another step-down—either to 2,300 to 4,000 V. In this way, the current is increased beyond its initial value at the generating station.

Finally, in the immediate area of the residential power consumer, the voltage is further stepped-down to 115/230 V, at which value it is used. The method of obtaining those two voltages, 115 V and 230 V, from a single source is explained in detail later in Sec. 2.5.

By means of transformers the utility company generates both the high voltage and relatively low current needed to reduce transmission-line losses and the relatively low voltage and high current needed by the residential consumer.

2.2   *Alternating Current*

Alternating-current power sources are called either *ac generators* or *alternators* and combine physical motion and magnetism to produce ac voltage. In Sec. 1.18 you learned that when a conductor passes through a magnetic field and cuts the lines of force (flux) an EMF is induced in the conductor. This is the basic principle on which an alternator operates.

The simplest form of alternator is shown in Fig. 2-1. To make it, a single loop of wire is placed between the poles of a permanent magnet and then made to rotate by some suitable mechanical device. As this loop rotates it cuts the magnetic lines of force and an EMF is induced in the loop. The EMF produced exists between the two ends of the loop. *Slip rings* (one attached to each end of the loop) and *brushes*, in physical contact with the slip rings, apply the generated EMF to an external circuit.

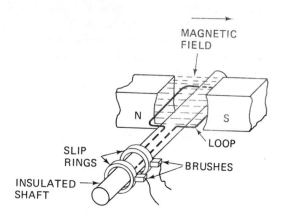

Figure 2-1   The simplest form of alternator

Refer to Fig. 2-2. When the loop is parallel to the magnetic lines of force, no flux is cut and no EMF is induced in the loop.

As the loop begins to rotate, however, it cuts the magnetic flux at an increasing rate; therefore, the induced EMF becomes steadily larger. When the loop has rotated one-quarter turn (90°) from its starting position, the maximum EMF is induced in the loop because it is now at right angles to the magnetic field. That is to say, it is cutting the maximum number of lines of force.

As rotation continues past the one-quarter point, the induced EMF is still in the *same* direction, but it is decreasing in value. At the half-revolution point (180°), the loop is again parallel to the magnetic field and the induced EMF is zero. The loop has now made a one-half turn, during which the induced EMF gradually increased to its maximum and then gradually fell to zero. *Since the sides of the loop have now reversed position, the induced EMF reverses polarity.* The EMF, however, again increases to a maximum at the three-quarter point (270°) of rotation; then it declines to zero when the rotation is completed (360°). The action described is repeated during each rotation.

The *wave form* of the induced EMF is often referred to either as a *sine-wave* or a *sinusoidal* wave. An alternating current has the same general wave form. Notice that the portion of the wave form *above* the horizontal axis is considered to be *positive*, while that lying *below* the horizontal axis is considered to be *negative*.

Each complete revolution of the loop is called one *cycle*. The number of cycles completed in one second (c/s), now ordinarily designated as hertz, identifies the *frequency* of the voltage or current. Abbreviated Hz, the term

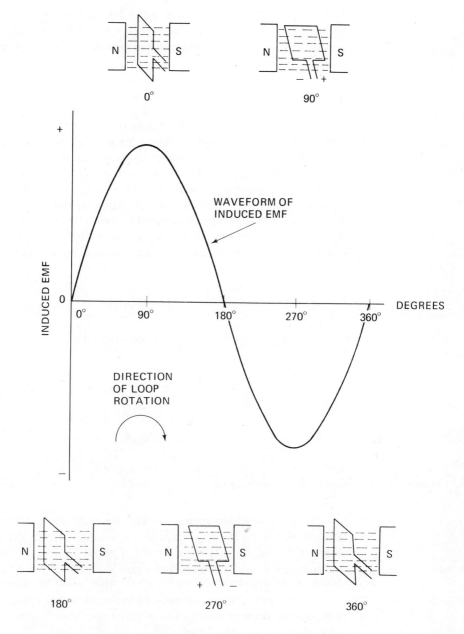

Figure 2-2 Generation of a sine-wave voltage

honors the memory of Heinrich Hertz, an early pioneer in the study of electrical energy.

Most electric power generated in the United States has a frequency of

60 Hz, but there is still a limited use of 50 Hz. All appliances, both small and heavy, are marked for use at 50/60 Hz. Remember that operation at frequencies outside this range may either damage or destroy an appliance or make it operate improperly.

Although the wave form in Fig. 2-2 is plotted on a horizontal axis marked in *degrees*, you will also find many wave-form representations where the horizontal axis is marked in terms of *elapsed time* in *seconds*.

Now, if two simple alternators start at precisely the same instant and rotate in the same direction at exactly the same speed, wave forms representing the EMFs induced in their loops would start and end at the same points on a common graph, such as that shown in Fig. 2-3. In addition, the maximum positive, the maximum negative, and the zero points would all correspond in time. The voltages or currents the waveshapes represent are then said to be *in phase*. Thus, the term phase indicates the *time* relationship between alternating voltages and/or currents. Notice, in particular, that the relative amplitudes or sizes of the wave forms are of no consequence in determining phase.

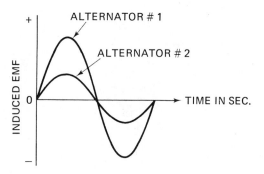

Figure 2-3   Two in phase waveshapes

If the same two alternators start their rotation at different times or in different directions, or if their speeds of rotation are not identical, the wave shapes of the induced EMFs will be *out of phase*. The *phase difference* between the two wave forms is expressed in *degrees*. Figures 2-4(a), (b), and (c) show wave forms having phase differences of 90°, 180°, and 270° respectively.

The terms *lead* and *lag* are used to describe the relative positions in time of voltages or currents that are out of phase. The quantity that is ahead in time *leads*, while the one behind *lags*.

2.3   *The Effective Value of AC*

As you know, dc remains at a constant level at all times. Thus, when a direct current of, say, 1 A passes through a fixed resistance, it produces a definite

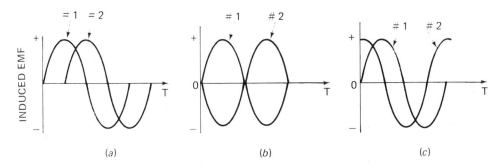

Figure 2-4   Wave forms having phase differences of (a) 90°, (b) 180°, and (c) 270°

amount of heat ($P = I^2R$). An alternating current having a *maximum* (*peak*) value of 1 A, in passing through the same resistance., would not produce the same amount of heat, however, since its value would be 1 A for only a very small part of the cycle and less than 1 A during the rest of the cycle. To compare dc and ac in terms of *heating effect,* ac is expressed in terms of its so-called *effective* value. By definition, *the effective value of an alternating current or voltage is that value of the quantity which produces the same heating effect in a resistance as an equal value of dc.* Mathematically, it can be shown that of effective value of an alternating current or voltage is equal to 0.707 or 70.7% of the peak (maximum positive or negative) value.

The 115/230 V ac found in residences is an effective value; that is, it will produce just as much heat in a resistance as 115/230 V dc. Moreover, the scale of instruments you may use to measure ac voltage or current will have at least one scale calibrated in effective values.

Sometimes, the abbreviation rms, meaning *root-mean-square* and indicating the procedure for its mathematical derivation, is used in place of the term "effective." Just remember that rms and effective refer to one and the same thing.

2.4   *Polyphase AC*

The simple alternator that you considered in Sec. 2.2 contained a single loop which rotated in a magnetic field produced by a permanent magnet. In practical alternators, that magnetic field is produced by an electromagnet. Since a steady magnetic field is required, dc is sent through the electromagnet. In place of the single loop, many turns of insulated wire are wound on an iron core. The whole assembly is referred to as the *armature*. See Fig. 2-5.

Suppose two separate armature coils are wound upon the same core. If the coils are wound one over the other, are electrically separate, and each has its own set of slip rings and brushes, ac will be induced in each coil as the armature

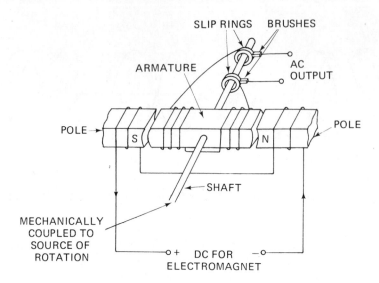

Figure 2-5   Simplified illustration of a practical alternator

rotates. Because the coils are wound over each other and rotate together, their induced voltages will be in phase.

Now suppose that, instead of having the coils wound one over the other, they are wound at right angles to each other, as shown in Fig. 2-6. At the instant armature coil A is cutting the maximum lines of force and has its maximum induced EMF, coil B is not cutting any lines and its induced voltage is zero. Because the two coils are at right angles to each other (90° apart), the phase difference between their induced EMFs is 90°.

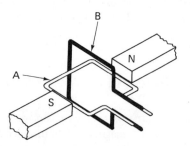

Figure 2-6   Two armature coils at right angles to each other

Going one step further, if three coils, each spaced 120° from its neighbor, are wound on the same core, the phase difference between the three induced voltages is 120°. These relationships are shown graphically in Fig. 2-7. Figure 2-7(a) shows the graph of the ac voltage produced by the alternator with

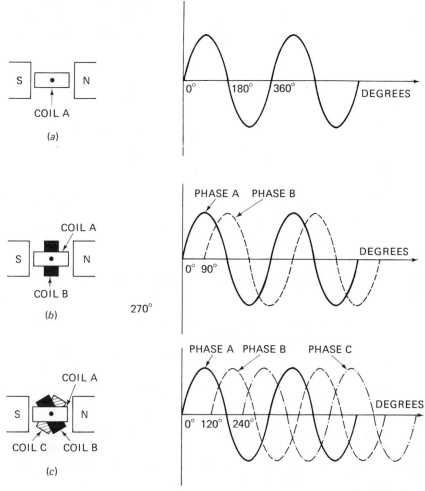

Figure 2-7   Alternators and the wave forms of their induced EMFs: (a) single-phase, (b) two-phase, and (c) three-phase

a single armature coil. Such an alternator is called a *single-phase* alternator. Figure 2-7(b) shows the graph of the voltages produced by a *two-phase* alternator in which the two windings are spaced 90° apart. The EMF of phase B starts 90° behind that of phase A and this time relationship is maintained at all times. Figure 2-7(c) shows the graph of the voltages produced by a *three-phase* alternator; that is, an alternator with three armature coils spaced 120° apart. The induced EMF of phase B lags behind that of phase A by 120° and the induced EMF of phase C lags 120° behind that of phase B.

      Alternators that have more than one set of armature coils are called *polyphase* alternators and the ac voltages and currents they produce are generally referred to as *polyphase* voltages and currents.

Most residential lighting systems and small appliances are constructed to operate on 120 V, 60 Hz, single-phase ac. (Actually, the electricity may be supplied at a voltage ranging anywhere between roughly 110 to 125 V, depending on many factors, but this is of no real consequence.) Electric ranges, water heaters, and some heavy appliances, such as large air-conditioning units, operate on 230 V, 60 Hz, single-phase ac. Both voltages are supplied to the individual residences by means of what is called the *three-wire* system which is explained in Sec. 2.6.

Two-phase systems are used in so few locations they do not merit further consideration.

Three-phase ac is used in power distribution systems and in commercial and industrial applications. Although three-phase ac is rarely encountered in residential wiring, some understanding of three-phase systems, such as how single-phase ac is derived from a three-phase system and how power is measured in a three-phase system, should prove beneficial. These subjects are examined, therefore, in Secs. 2.6 and 2.7.

2.5   *Single-Phase Transformers*

A simple *single-phase transformer* has two coils of wire which are *magnetically* coupled to, but *electrically* isolated from, each other. One coil, called the *primary,* is connected to an ac source of EMF, and the other coil, called the *secondary*, is connected to the load. When ac flows through the primary, the varying magnetic field of the primary cuts across the turns of the secondary and induces an EMF therein.

In an *ideal* or theoretically perfect transformer, all lines of force around the primary *link* or cut across the turns of the secondary. Under these conditions, there is 100 percent magnetic *coupling*. In a practical device, however, some of the magnetic lines of force leak off into the air and are called *leakage flux*.

To reduce leakage flux, the primary and secondary windings are wound on a magnetic core which, because it has a much higher permeability than air, concentrates the lines of force and forms a closed magnetic circuit.

In a *shell-core* transformer, Fig. 2-8, the primary and secondary windings are placed one over the other on the center arm of the core. This type of transformer, which you will encounter often, is generally designed to produce a magnetic coupling that approaches 100 percent effectiveness.

In addition to losses resulting from leakage flux, a transformer has *copper* and *iron* losses. The copper loss is due to the resistance of the wire used in the windings. The iron losses result from *hysteresis* and *eddy currents*. Hysteresis loss is the loss of energy required to make the core change the polarity of its poles in step with the frequency of the ac. This loss is reduced by the use of alloys which have a greater permeability than iron. Eddy currents are small circulating currents induced in the core by the changing magnetic field and

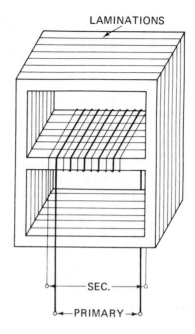

Figure 2-8   A shell-core transformer—the thickness of the laminations is grossly exaggerated

are reduced by making the core out of very thin strips called *laminations*. Each is coated with an insulation of varnish so that eddy currents cannot circulate through the core.

Figure 2-9 shows the schematic diagram of a transformer connected to a source and a load. Note the symbol for an ac source of EMF.

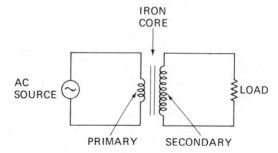

Figure 2-9   Schematic diagram of a transformer connected to a source and load

If the secondary circuit is open at some point, only a small current flows through the primary since the magnetic field, constantly cutting across the turns of the primary, induces in it a CEMF that is nearly equal to the applied EMF. The small amount of power consumed is used mainly to magnetize the core.

Notice, incidentally, that if dc were applied to the primary, no CEMF would be developed and current would be limited only by the low resistance of the winding. The resulting high current would burn out the primary. Thus, a transformer is *never* used in a dc circuit.

Now, if the secondary circuit is closed, current flows through the secondary winding and load. As we know from Lenz's law, the magnetic field around the secondary tends to neutralize a portion of the magnetic field around the primary. Thus, the CEMF induced in the primary decreases and the primary current increases.

As the load is increased, more current flows through the secondary circuit, the induced CEMF in the primary is decreased still more and the primary draws more current from the source. As the load is decreased, however, less current flows through the secondary winding, the CEMF induced in the primary increases and less current is drawn from the source. This shows us that, within limits, the transformer automatically adjusts itself to changes in the load. If the load is too great, however, the resulting high current through the primary circuit will burn out the primary winding.

To demonstrate another useful function of the transformer, assume an ideal transformer with 100 percent magnetic coupling and having a primary winding of 120 turns connected to a 120 V source of EMF.

When the secondary circuit is open, it consumes no power. Since, in the ideal transformer, the power consumed in the primary and secondary circuits are equal, no power is consumed in the primary circuit either.

The CEMF produced by the primary is equal to the voltage of the source. As the magnetic field cuts across the 120-turn primary, therefore, the EMF induced in each turn is one volt. Since the same magnetic field cuts across the turns of the secondary, each turn of the secondary also has one volt induced in it. If the secondary winding also has 120 turns, the induced voltage across the secondary is 120 V.

Suppose, however, that there are only 12 turns on the secondary. With one volt induced in each turn, the secondary voltage is then 12 V. This arrangement is called a *step-down* transformer. If the secondary contains 1200 turns, the secondary voltage is 1200 V and that arrangement is called a *step-up* transformer. Thus, as noted in Sec. 2.1, a transformer can be used to either step-up or step-down alternating voltage.

The above considerations show us that the ratio of the voltages across the transformer windings is equal to the turns ratio of the windings. This relationship is expressed by the formula

$$\frac{E_p}{E_s} = \frac{N_p}{N_s} \qquad (2\text{-}1)$$

where E_p is the voltage across the primary, E_s the voltage across the secondary,

N_p the number of turns on the primary, and N_s the number of turns on the secondary.

If, for example, N_p = 500 turns, N_s = 5,000 turns, and E_p = 120 V,

$$E_s = \frac{E_p N_s}{N_p} = \frac{120 \times 5000}{500} = 1200 \text{ V}$$

Again, in an ideal transformer, we assume no losses in our theoretical calculating. Thus, the power of the secondary circuit ($E_s I_s$) is equal to the power of the primary circuit ($E_p I_p$). If, as in the above example, E_s is stepped-up ten times, I_s will be reduced to one-tenth that of the primary. This relationship is given in the following formula:

$$\frac{I_p}{I_s} = \frac{N_s}{N_p} \tag{2-2}$$

Using the same values for N_p and N_s as in the preceding example, and letting I_p = 1 A,

$$I_s = \frac{I_p N_p}{N_s} = \frac{1 \times 500}{5000} = \frac{1}{10} = 0.1 \text{ A}$$

So we see that a transformer can also be used to step-up or step-down alternating current. Of course, if the voltage is stepped-up the current is stepped-down and vice versa.

Transformers may be constructed with two or more secondary windings to obtain both step-up and step-down relationships.

A variation of the two-winding transformer is the *autotransformer* illustrated in Fig. 2-10. This transformer has a single *tapped* coil. The turns between the tap and the two ends form the primary and secondary windings. A step-up arrangement is shown in Fig. 2-10(a) and a step-down arrangement in

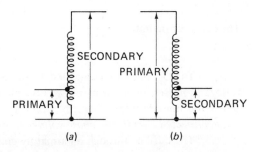

(a)   (b)

Figure 2-10   Autotransformer: (a) step-up and (b) step-down

Fig. 2-10(b). Except in a few special applications, however, such as in connection with motor starting devices, the use of autotransformers is prohibited by the National Electrical Code. [See NEC Sections 210-9, 410-78, and 430-82(b).]

2.6　Three-Phase Machines and Transformers

Where large quantities of electric power are needed, three-phase ac, usually at 60 Hz, is supplied to the power mains. As noted in Sec. 2.4, such currents are generated by an alternator having three identical armature coils spaced 120° from each other. Since each armature coil has two ends, seemingly, six lines (two for each coil) are needed to transfer current to the load. Fortunately, it is possible to join one end of each coil at the alternator and transmit current over three lines, one for each phase. Two configurations, called the *delta* and *wye*, can be used to connect the armature coils.

Delta configuration

In the delta configuration, Fig. 2-11, the ends of each armature coil are joined to the ends of the other two coils. With this arrangement, the *net* voltage around the triangle or delta (Δ) formed by the coils is zero. Of course, this means that current around the triangle is also zero.

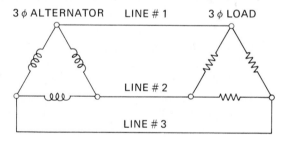

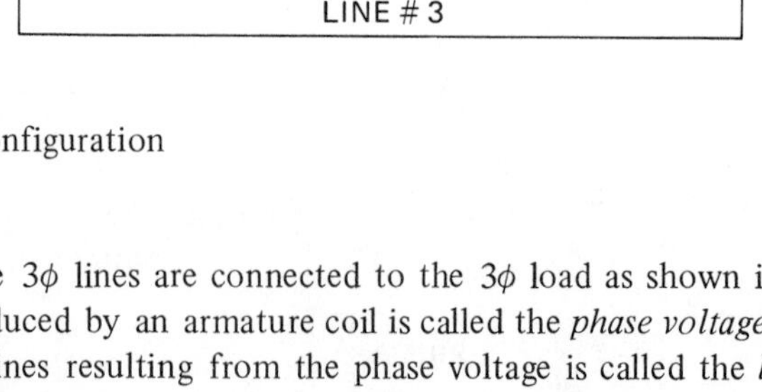

Figure 2-11　　The delta configuration

The 3ϕ lines are connected to the 3ϕ load as shown in Fig. 2-11. The voltage produced by an armature coil is called the *phase voltage*, and the voltage across the lines resulting from the phase voltage is called the *line voltage*. *In a delta-connected circuit, the phase voltage is the same as the line voltage.*

The current through an armature coil, when it is connected to a load, is called the *phase current*, and the corresponding currents flowing through the lines is called the line current. *In the delta configuration, line current is equal to*

1.73 times the phase current. At any one instant, current flows from the alternator through one line to the load and returns via the other two lines, or conversely the current flows from the alternator through two lines and returns by way of the third.

In Fig. 2-11 the load on the line is assumed to be *balanced*; that is, the load is such that all line currents are equal and have the same phase angle with respect to line voltages. Power companies always try to maintain a balanced condition because it reduces power losses to a minimum. A balanced load is, therefore, assumed in all further discussion in this book.

To step-up or step-down 3ϕ ac, three identical single-phase transformers, one for each phase, can be used. See Fig. 2-12(a). Notice that all primary and secondary windings are delta connected. Because the complexities of the diagram in Fig. 2-12(a) tend to be confusing, the usual schematic representation of a 3ϕ transformer appears in the detail shown in Fig. 2-12(b). In place of three single-phase transformers, a single three-phase transformer, with all windings on a single core, may be used. For either case, however, the schematic representation is the same.

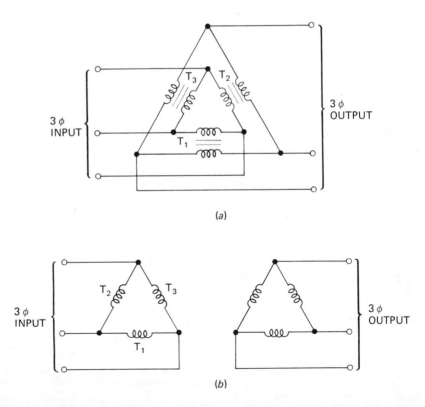

(a)

(b)

Figure 2-12   (a) Three identical single-phase transformers with all primary and secondary windings delta connected, and (b) usual schematic representation of same

To obtain 230 V single-phase ac from a delta-type secondary, connections are made across any *one* phase of the secondary. In Fig. 2-13, phase *C* has been selected for this purpose. To obtain the 115 V needed for lighting circuits and small appliances, a *neutral* wire is connected to a center tap on phase *C*. In practically all modern installations the neutral wire is grounded. Assuming a grounded-neutral system, 230 V is applied to any device connected across the entire single-phase winding, while 115 V is delivered to any device connected between the neutral wire and either of the two outer (*hot*) wires. In actual wiring, the neutral line is *white*, and the *hot* wires are usually *black* (but *never* white or green).

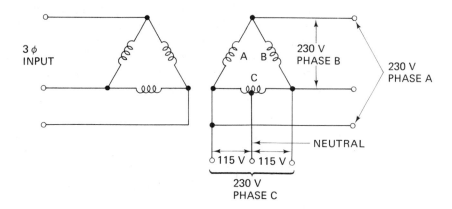

Figure 2-13   Obtaining 115/230 V single-phase ac from a delta-type transformer secondary

Notice that 3φ ac at 230 V is also obtainable from the same secondary by connecting to all three wires. In Fig. 2-13 the power is delivered at 230 V and this is the usual arrangement for residences. For other applications, the power may be delivered at either 460 or 575 volts. The voltage depends, of course, on the transformers selected for connection to the power company supply lines.

Wye configuration

In this Y-shaped arrangement one end of each armature coil is connected to one end of both the others. See Fig. 2-14. The free ends of the coils are joined to the 3φ lines. The coils must be connected with correct polarities, as indicated. *In the wye configuration, the line current is the same as the phase current, but the line voltage is 1.73 times the phase voltage.*

As in the delta configuration, single-phase ac is obtained across any two of the three lines.

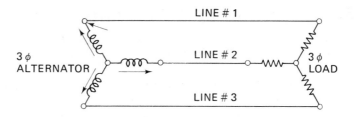

Figure 2-14   The wye configuration

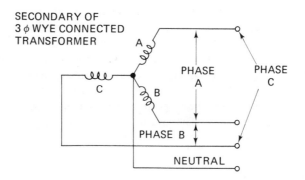

Figure 2-15   The three-phase, four-wire system

A variation of the wye configuration is shown in Fig. 2-15. Known as a three-phase, *four-wire system*, it has a grounded neutral connected to the junction of the transformer's secondary coils. This modified configuration offers certain advantages.

In a three-wire system, 3ϕ ac is obtained by connecting a load to all three wires. When single-phase ac is desired, the load is connected to any two of the wires. In the four-wire system, single-phase ac is obtained by connecting the single-phase load to the neutral and any one of the three *hot* wires.

Also in the four-wire system, if the single-phase current is obtained by connecting to the neutral and to one of the other three wires *and* if the 3ϕ voltage is 208 V (the usual case), the single-phase voltage will be 120 V. Thus, by using four wires we can obtain either a 3ϕ current at 208 V for water heaters and similar large loads, or a 3ϕ current at 230 V for 3ϕ motors, or a single-phase current at 120 V for lighting and small appliances.

In conclusion, it is worth noting that for the same line voltage and current a three-wire, 3ϕ system will deliver 1.73 times as much power as a two-wire, single-phase system. Thus, less copper is needed for the transmission of 3ϕ power than for an equivalent single-phase power.

2.7   *Power in AC Circuits*

Reactive power of an inductance

Power supplied to an inductance to build up its magnetic field is stored as potential energy in the field itself. When the field collapses, this energy is *returned* to the circuit by induction. Since the inductance in an ac circuit alternately takes from and returns to the source equal amounts of energy, *no true power is dissipated by an inductor*.

In the case of a pure resistance, we are able to determine the true power by simply multiplying the effective root-mean-square values (rms) of voltage and current. In the circuit of Fig. 2-16, even though the meters indicate the effective values of voltage and current for the pure inductance, the product of V and *I* does not represent true power since, as we have just seen, the true power is zero. The V*I* product is, however, directly proportional to the amount of energy stored and returned by the inductor each time the current changes direction; therefore, the V*I* product in a pure inductance is called the *reactive power* of the inductance. The unit of reactive power is the *volt-ampere (reactive)*, abbreviated *var*.

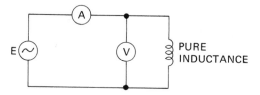

Figure 2-16   The reactive power of an inductance

Reactive power of a capacitance

In working with electric motors, electricians often encounter a device called a *capacitor*. (Condenser is the obsolete name for capacitor.) *A capacitor is simply a device that stores an electric charge under pressure. The ability of a capacitor to store an electric charge is called capacitance.*

In physical form, a capacitor is simply two parallel conducting plates separated by an insulating material known as a *dielectric*. Small capacitors are often constructed as shown in Fig. 2-17. Here, two long strips of aluminum foil are rolled up with two strips of wax-paper dielectric. Leads are attached to each strip of foil for connection into a circuit and the entire assembly is sealed in a cardboard casing. A capacitor of this type is referred to as a *lumped* capacitor since its total capacitance is lumped into a single component.

In electric circuits *distributed* capacitance refers to that which is continuous along the whole length of a line or is the set of capacitances between

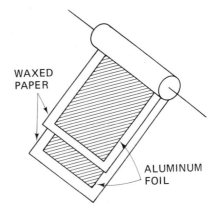

WAXED
PAPER

ALUMINUM
FOIL

Figure 2-17   Typical construction of a small capacitor

various parts of a coil. In the armature of an electric motor, for example, there are many parallel conductors separated by insulation. That is, the motor possesses distributed capacitance.

In the same way that a magnetic field is associated with an inductance, an *electric field* is associated with a capacitance—either lumped or distributed. *This field exists between the parallel conductors.*

Power supplied to a capacitor to build up its electric field is stored as electric energy in the field itself. If the direction of current reverses, this energy is *returned* to the circuit. Since the capacitances in an ac circuit alternately take from and return to the source equal amounts of energy, *no true power is dissipated by a capacitance.*

Again, the product of V and *I* in a pure capacitance does not represent true power; instead, V*I* represents the reactive power of a capacitance and is expressed in vars.

Apparent power

If we use a voltmeter and an ammeter to measure the total voltage and total current in an ac circuit, their product represents neither the true power in watts nor the reactive power in vars; instead, the product is called the *apparent power.* Since apparent power is simply the product of volts and amperes, the unit of apparent power is the *volt-ampere* abbreviated VA.

Power factor

The apparent power that a source must supply to a *reactive* load (one containing inductance and capacitance) is always greater than the true power which the load converts into some other form of energy.

The ratio between the true power (in watts) and the apparent power (in volt-amperes) of a load in an ac circuit is called the *power factor* of the load. From this definition,

$$\text{Power factor} = \frac{\text{Watts}}{\text{Volt-amperes}} \qquad (2\text{-}3)$$

Power factor is often abbreviated *PF*. Since it is a simple ratio, power factor is a positive decimal fraction, which may be changed to a percentage simply by shifting the decimal point two places to the right.

To distinguish between inductive and capacitive loads, inductive loads are considered to have a *lagging* power factor and capacitive loads a *leading* power factor.

Although in residential wiring an electrician is not concerned with power factor, it is a very important consideration in installations, such as those on a farm where a number of *induction* motors and transformers may be in operation. Since that type of load is inductive, the power factor will be lagging.

A voltmeter, ammeter, and wattmeter are used to determine power factor.

Suppose, for example, a single-phase induction motor is connected to a 230 V source and draws 80 A. The wattmeter reads 12.5 kW. The power factor is

$$PF = \frac{12{,}500}{230 \times 80} = \frac{12{,}500}{18{,}400} = 0.69 = 69\% \text{ lagging}$$

If a motor is operated from a 3ϕ circuit, multiply the VA by 1.73. If, for example, a 3ϕ motor connected to a 230 V source draws 14 A and the wattmeter indicates 4200 W,

$$PF = \frac{4200}{1.73(230 \times 14)} = \frac{4200}{5571} = 0.75 = 75\% \text{ lagging}$$

In most cases, power factor improves as the horsepower of the motor is increased. Typical power factors range from a low of about 50 percent for small fractional-horsepower motors to a high of about 92 percent for 25-horsepower motors.

Importance of high-power factor

Since reactive power, which does no useful work, must be supplied by the power source, the source must be large enough to furnish both the reactive power and the true power which does useful work. Thus, larger and heavier equipment and conductors are needed to provide a given amount of useful power to a load having a low-power factor than to one having a relatively high-power factor. Since it costs electric suppliers more to deliver such power, they charge the user a higher rate to compensate for their investment.

Conductors in the consumer's installation must also be larger than necessary to safely convey both the true power and reactive power. Heavier than necessary conductors can represent a substantial and needless investment, however.

Power-factor correction

The cheapest and most widely used method for correcting the lagging-power factor caused by inductive loads is to introduce a leading-power factor into the circuit. This is accomplished by connecting a *fixed* capacitor in *parallel* with the load.

Table 2-1 shows how to calculate the size capacitor needed to obtain an *improved* power factor. Suppose, for example, we want to improve a lagging-power factor from 72 percent to 90 percent. Present power factors are listed in the right-hand column and desired power factors across the top of Table 2-1. Entering the table at 72 percent and moving to the right, we find the factor 0.48 in the column headed 90 percent. When a true power is multiplied by 0.48 the product is the *kilovar* (kvar) of reactive power needed to raise the power factor to 90 percent. If the true power is, for example, 12,500 W, the needed reactive power is

$$12{,}500 \times 0.48 = 6000 \text{ var} = 6 \text{ kvar}$$

Capacitors for this purpose come in standard units whose name plates list, among other factors, the rated voltage and frequency at which the unit is to be operated and the reactive power in kvar that it will introduce into the circuit. A standard unit having a kvar rating closest to the calculated value is connected in parallel with the load. If necessary, several standard capacitors may be hooked in parallel to obtain a value equal to the *sum* of the kvar rating of the individual units.

Although a 100 percent power factor is the best theoretical solution, the cost of achieving it must be weighed against the practical cost of operating at a somewhat lower power factor. The capacitors use oil as a dielectric, are hermetically sealed in a steel can, and are expensive.

Since a capacitor stores an electric charge, there also is the danger of receiving a *severe* shock when one is handled. To minimize this danger, the National Electrical Code requires that internal resistors or inductors be included so as to discharge the capacitor within a stated interval of time. Once this time has elapsed, the capacitor may be removed from the line without danger of shock.

In 3ϕ circuits, each phase must be corrected. A three-phase capacitor,

resultant or desired power factor in percent

present power factor	80	81	82	83	84	85	86	87	88	89	90	91	92	93	94	95	96	97	98	99	100
50	.98	1.01	1.03	1.06	1.09	1.11	1.14	1.16	1.19	1.22	1.25	1.28	1.30	1.34	1.37	1.40	1.44	1.48	1.53	1.59	1.73
51	.94	.96	.99	1.01	1.04	1.07	1.09	1.12	1.15	1.17	1.20	1.23	1.26	1.29	1.32	1.36	1.39	1.43	1.48	1.54	1.69
52	.89	.92	.95	.97	1.00	1.02	1.05	1.08	1.10	1.13	1.16	1.19	1.21	1.25	1.28	1.31	1.35	1.39	1.44	1.50	1.65
53	.85	.88	.90	.93	.95	.98	1.01	1.03	1.06	1.09	1.12	1.14	1.17	1.20	1.24	1.27	1.31	1.35	1.40	1.46	1.60
54	.81	.83	.86	.89	.91	.94	.97	.99	1.02	1.05	1.07	1.10	1.13	1.16	1.20	1.23	1.27	1.31	1.36	1.42	1.56
55	.77	.79	.82	.85	.87	.90	.93	.95	.98	1.01	1.03	1.06	1.09	1.12	1.16	1.19	1.23	1.27	1.32	1.38	1.52
56	.73	.76	.78	.81	.83	.86	.89	.91	.94	.97	1.00	1.02	1.05	1.08	1.12	1.15	1.19	1.23	1.28	1.34	1.48
57	.69	.72	.74	.77	.80	.82	.85	.87	.90	.93	.96	.99	1.01	1.05	1.08	1.11	1.15	1.19	1.24	1.30	1.44
58	.65	.68	.71	.73	.76	.78	.81	.84	.86	.89	.92	.95	.98	1.01	1.04	1.08	1.11	1.15	1.20	1.26	1.40
59	.62	.64	.67	.70	.72	.75	.77	.80	.83	.86	.88	.91	.94	.97	1.00	1.04	1.08	1.12	1.16	1.23	1.37
60	.58	.61	.64	.66	.69	.71	.74	.77	.79	.82	.85	.88	.90	.94	.97	1.00	1.04	1.08	1.13	1.19	1.33
61	.55	.57	.60	.63	.65	.68	.71	.73	.76	.79	.81	.84	.87	.90	.94	.97	1.01	1.05	1.10	1.16	1.30
62	.51	.54	.57	.59	.62	.64	.67	.70	.72	.75	.78	.81	.84	.87	.90	.94	.97	1.01	1.06	1.12	1.26
63	.48	.51	.53	.56	.59	.61	.64	.67	.69	.72	.75	.78	.80	.84	.87	.90	.94	.98	1.03	1.09	1.23
64	.45	.48	.50	.53	.55	.58	.61	.63	.66	.69	.72	.74	.77	.80	.84	.87	.91	.95	1.00	1.06	1.20
65	.42	.44	.47	.50	.52	.55	.58	.60	.63	.66	.68	.71	.74	.77	.81	.84	.88	.92	.97	1.03	1.17
66	.39	.41	.44	.47	.49	.52	.54	.57	.60	.63	.65	.68	.71	.74	.77	.81	.85	.89	.93	1.00	1.14
67	.36	.38	.41	.44	.46	.49	.51	.54	.57	.60	.62	.65	.68	.71	.74	.78	.82	.86	.90	.97	1.11
68	.33	.35	.38	.41	.43	.46	.49	.51	.54	.57	.59	.62	.65	.68	.72	.75	.79	.83	.88	.94	1.08
69	.30	.32	.35	.38	.40	.43	.45	.48	.51	.54	.56	.59	.62	.65	.69	.72	.76	.80	.84	.91	1.05
70	.27	.30	.32	.35	.37	.40	.43	.45	.48	.51	.54	.56	.59	.62	.66	.69	.73	.77	.81	.88	1.02
71	.24	.27	.29	.32	.35	.37	.40	.42	.45	.48	.51	.54	.56	.60	.63	.66	.70	.74	.78	.85	.99
72	.21	.24	.26	.29	.32	.34	.37	.40	.42	.45	.48	.51	.53	.57	.60	.63	.67	.71	.75	.82	.96
73	.19	.21	.24	.26	.29	.32	.34	.37	.40	.42	.45	.48	.51	.54	.57	.61	.64	.68	.73	.79	.94
74	.16	.18	.21	.24	.26	.29	.32	.34	.37	.40	.42	.45	.48	.51	.55	.58	.62	.66	.70	.77	.91
75	.13	.16	.18	.21	.24	.26	.29	.31	.34	.37	.40	.43	.45	.49	.52	.55	.59	.63	.67	.74	.88
76	.10	.13	.16	.18	.21	.23	.26	.29	.31	.34	.37	.40	.43	.46	.49	.53	.56	.60	.65	.71	.85
77	.08	.10	.13	.16	.18	.21	.24	.26	.29	.32	.34	.37	.40	.43	.47	.50	.54	.58	.62	.69	.83
78	.05	.08	.10	.13	.16	.18	.21	.24	.26	.29	.32	.35	.37	.41	.44	.47	.51	.55	.59	.66	.80
79	.03	.05	.08	.10	.13	.16	.18	.21	.24	.26	.29	.32	.35	.38	.41	.45	.48	.52	.57	.63	.78
80	.00	.03	.05	.08	.10	.13	.16	.18	.21	.24	.27	.29	.32	.35	.39	.42	.46	.50	.54	.61	.75
81		.00	.03	.05	.08	.10	.13	.16	.18	.21	.24	.27	.29	.33	.36	.39	.43	.47	.51	.58	.72
82			.00	.03	.05	.08	.10	.13	.16	.19	.21	.24	.27	.30	.33	.37	.41	.45	.49	.56	.70
83				.00	.03	.05	.08	.10	.13	.16	.19	.22	.24	.28	.31	.34	.38	.42	.46	.53	.67
84					.00	.03	.05	.08	.10	.13	.16	.19	.22	.25	.28	.32	.35	.39	.43	.50	.65
85						.00	.03	.05	.08	.11	.14	.16	.19	.22	.26	.29	.33	.37	.42	.48	.62
86							.00	.03	.05	.08	.11	.14	.17	.20	.23	.26	.30	.34	.39	.45	.59
87								.00	.03	.05	.08	.11	.14	.17	.20	.24	.27	.32	.36	.42	.57
88									.00	.03	.06	.08	.11	.14	.18	.21	.25	.29	.34	.40	.54
89										.00	.03	.05	.09	.12	.15	.18	.22	.26	.31	.37	.51
90											.00	.03	.06	.09	.12	.15	.19	.23	.28	.34	.48
91												.00	.03	.06	.09	.13	.16	.21	.25	.31	.46
92													.00	.03	.06	.10	.13	.18	.22	.28	.43
93														.00	.03	.07	.10	.14	.19	.25	.39
94															.00	.03	.07	.11	.15	.22	.36
95																.00	.04	.08	.12	.19	.33
96																	.00	.04	.09	.15	.29
97																		.00	.05	.11	.25
98																			.00	.06	.20
99																				.00	.14
100																					.00

Table 2-1   Factors for calculating the size capacitor needed to obtain an improved power factor with an inductive load (Westinghouse Electric Corp.)

having three capacitors connected in either a delta or wye configuration, is contained in a single case. Separate connectors are provided on top of the case to connect one capacitor to each line.

Either of two methods can be used in connecting capacitors for power-factor correction. One method is to calculate the size capacitor needed to correct the power factor for the complete installation and then to connect this capacitor to the main power bus near the *service entrance*. (The general term "service entrance" includes all wires and equipment from the outside of the building up to and including the meter; a "disconnect" means to disconnect the entire installation from the power supplier's lines and overcurrent protection, such as circuit breakers or fuses, and from the ground.) Since the capacitor corrects the power factor from the point of insertion *back* to the source, the correction is not applied on lines leading to individual appliances. This method reduces the penalty imposed by the electric supplier for a low-power factor.

A second method is to connect a capacitor at each appliance. Although this method also corrects the low-power factor on individual feeder lines inside the plant, it is usually more costly than the method described above. The choice of method is usually dictated by economics.

Measuring power in three-phase circuits

When two wattmeters are connected as shown in Figs. 2-18(a) and (b), the total power in either a balanced or unbalanced load is the *algebraic sum* of the two meter readings.

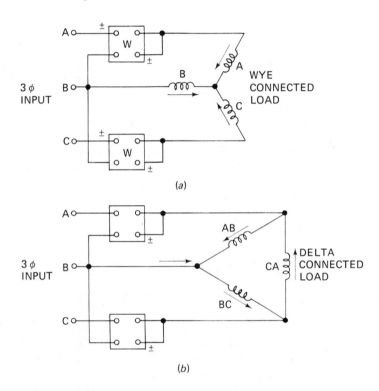

(a)

(b)

Figure 2-18   Measuring power in 3ϕ (a) wye and (b) delta systems

EXERCISES

2-1/   What is an *alternating* current?

2-2/   Which of two possible methods is ordinarily used to reduce power losses on electric transmission lines? Why?

2-3/   What is the basic principle on which an alternator works?

2-4/   Describe the operation of a simple (single-loop) alternator.

2-5/   In connection with ac, what is meant by the term Hertz? Sixty Hz?

2-6/   When are two wave forms *in phase*? *Out of phase*?

2-7/   What is meant by the *effective* (rms) value of an alternating voltage or current?

2-8/   What is meant by the terms *polyphase voltage* and *polyphase current*?

2-9/   How are the armature coils of an alternator arranged to produce 3ϕ ac?

2-10/   Describe the physical arrangement of a single-phase transformer.

2-11/   What do we mean by the expression *ideal transformer*?

2-12/   What is *leakage flux*?

2-13/   Describe the physical arrangement of a *shell-core* transformer.

2-14/   In connection with transformer losses, what is meant by the terms *hysteresis* and *eddy currents*?

2-15/   In practical transformers, how are hysteresis and eddy current losses reduced?

2-16/   If the secondary circuit of a two-winding transformer is open, is the current passing through the primary circuit large or small? Why?

2-17/   Can a transformer be used in a dc circuit? Explain in detail.

2-18/   What happens to the primary current when the load on the secondary *increases*? Why?

2-19/   What happens to the primary current when the load on the secondary *decreases*? Why?

2-20/   What is a *step-up* transformer? A *step-down* transformer?

2-21/   What relationship exists between the *voltage* and *turns ratios* of a transformer primary and secondary?

2-22/   What relationship exists between the *current* and *turns ratios* of a transformer primary and secondary?

2-23/   What is an *autotransformer*? Are autotransformers widely employed in electric circuits? Explain in detail.

2-24/   In a *delta* configuration, what relationship exists between the *phase voltage* and the *line voltage*?

2-25/   In a *delta* configuration, what relationship exists between *line current* and *phase current*?

2-26/   How is 115/230 V single-phase ac obtained from a *delta*-type secondary?

2-27/   In a *wye* configuration, what relationship exists between the *line current* and *phase current*? Between the *line voltage* and *phase voltage*?

2-28/   What is a three-phase *four-wire* system?

2-29/   How is single-phase ac obtained from a 3ϕ four-wire system?

2-30/   Is any *true* power dissipated by an inductance? Explain in detail.

2-31/   What is the *unit* of reactive power?

2-32/   What is *capacitance*? A *capacitor*?

2-33/   *Where* is the electric charge stored in a capacitor?

2-34/   Does a capacitance dissipate any *true* power? Explain in detail.

2-35/   What is meant by the term *apparent power*? What is the *unit* of apparent power?

2-36/   What is meant by the term *power factor*?

2-37/   Describe the procedure for determining the power factor of the load in a single-phase system.

2-38/   Describe the procedure for determining the power factor of the load in a 3ϕ system.

2-39/   In what *two* ways can the power factor of a load be expressed?

2-40/   What types of load exhibit a *lagging* power factor? A *leading* power factor?

2-41/   Why is a *high* power factor important?

2-42/   What is the cheapest and most widely used method for improving the *lagging* power factor caused by inductive loads?

2-43/   Suppose you want to improve a lagging power factor from 70 percent to 92 percent. By what *factor* would you multiply true power to determine the reactive power required?

2-44/   Describe two methods for connecting capacitors to correct a lagging power factor.

2-45/   How can the power in a three-phase circuit be measured?

3

Illumination

3.1   *Basic Concepts*

In this text our previous references to lighting loads have been only in terms of power. Too many variables affect lighting, however, to say with certainty that a given amount of power dissipated will always provide a desired amount of illumination. Some of the variables affecting lighting effectiveness are the type of bulb used—incandescent or fluorescent—whether it is clear or frosted, how much surrounding areas reflect or diffuse the light, how efficiently the fixture is constructed for the concentration of illumination, and finally the distance from the lamp to the area where the level of lighting is measured.

Two interrelated units, the *footcandle* and the *lumen* are important in any consideration of illumination values.

In Fig. 3-1, a candle is located in the center of a sphere having a radius of one foot. Every point on the sphere is said to be illuminated to a level of one footcandle (fc) since each point is located one foot from the light source. Notice that the fc is a measure of illumination at a *point*; no area is involved.

When area is involved, the quantity of light is measured in lumens, abbreviated lm. One lumen is the amount of light falling on each square foot (ft²) of the sphere diagrammed in Fig. 3-1.

One lm of light is needed to produce a uniform illumination of 1 fc

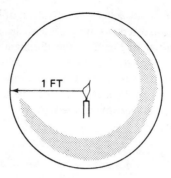

Figure 3-1   Defining the footcandle

over an area of 1 ft^2. To produce a level of illumination of, say, 10 fc, we must, therefore, provide 10 lm for every square-foot of area involved. Thus 1 fc = 1 lm/ft^2.

The intensity of illumination can be measured with a *footcandle meter* (commonly called a *light meter*) which is simply a *photoelectric cell* used in combination with a microammeter. When light falls on the cell, the cell generates an EMF which is proportional to the intensity of the light. Current produced by this EMF, in passing through the microammeter, deflects a pointer across a scale calibrated in fc so that readings can be made.

One other characteristic of light requires explanation at this point; namely, *light varies inversely as the square of distance from the light source*.

To illustrate this characteristic of light, suppose a hole with an area of 1 ft^2 is cut in the sphere of Fig. 3-1. With a light source of one candle in the

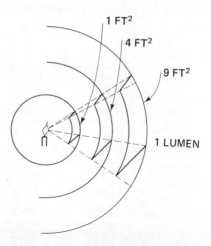

Figure 3-2   Light varies inversely as the square of distance from the light source

center of the sphere, exactly 1 lm of light will escape through the opening. Now, if rounded screens are placed at distances of 2 and 3 feet from the light source, as shown in Fig. 3-2, the illuminated areas of the screens measure 4 ft^2 and 9 ft^2, respectively. Since the amount of light remains the same (1 lm), the inner screen receives 1/4 lm/ft^2, while the outer screen receives only 1/9 lm/ft^2.

To determine the relative illumination of the outer screen to the inner screen, simply divide the square of the shorter distance by the square of the longer;

$$\frac{4}{9} = 44.4 \text{ percent.}$$

3.2   *Incandescent Lamps*

Incandescence means *a glowing due to heat*. Because *tungsten* has a very high melting point, it is the principal light source in incandescent lamps.

Incandescent lamps are reasonably inexpensive, have fair life expectancies, can be used in practically any environment, will continue to operate with wide variations in applied voltage, and are generally used with simple fixtures. Moreover, because their filaments are purely resistive, these lamps have a unity (1) power factor.

The incandescent lamps most often used in residential lighting range in size from about 25 W up to about 500 W and have standard *screw-shell* bases. The *mogul* is the largest of the screw-shell bases and is used mostly on lamps from 300 W upward.

The wattage rating, initial output in lumens, average output in lumens, and life expectancy in hours for typical incandescent lamps are shown in Table 3-1.

Rating in Watts	Initial output in lumens	Average output in lumens	Normal life expectancy in hours
25	265	230	1000
40	470	415	1000
60	840	795	1000
100	1640	1540	750
150	2700	2420	750
200	3800	3400	750
300	5750	5150	1000
500	9900	8750	1000

Table 3-1   Characteristics of incandescent lamps

As a lamp ages, its inner surface becomes blackened from minute tungsten deposits and the lumen output drops correspondingly as the tungsten filament burns. There is also a proportionately small decrease in power consumption. So, in areas where lighting levels are critical, replacement at regular intervals is recommended.

The lowest over-all cost of illumination is achieved by operating lamps at the voltage for which they are designed. Although operation at a lower voltage prolongs the life of a lamp bulb, the actual watts dissipated, the total output in lumens, and the output of lumens per watt show sharp declines. At higher than design voltages the opposite characteristics prevail. See Table 3-2.

Percent of rated voltage	Percent of Normal life expectancy	Percent of Actual watts dissipated	Percent of Total output in lumens	Percent of Output lumens per watt
85	800	78	58	72
90	400	85	68	80
95	200	93	83	90
100	100	100	100	100
105	58	108	118	109
110	37	116	140	120
115	18	124	162	131

Table 3-2   Voltage characteristics of incandescent lamps

3.3   *Fluorescent Lamps*

Fluorescent lamps offer several advantages over the incandescent variety: (1) high efficiency—with two to three times more lumens per watt (in color, up to about 100 times more); (2) less heat (an important consideration in air-conditioned spaces); (3) light with less glare or shadow which reduces eyestrain since it is distributed over a larger lamp-surface area; and (4) under ordinary operating conditions, five to ten times the life expectancy.

The disadvantages of fluorescent lamps as compared to their incandescent counterparts include: (1) temperature sensitivity (lamps intended for residential use operate at temperatures approaching 50°F; special lamps are available, however, for operation at lower temperatures); (2) high humidity may necessitate the use of a protective enclosure; (3) life expectancy is affected to a great extent by the number of times the lamp is turned on and off (published life expectancies assume the lamps operate continuously for three hours each time they are turned on); (4) a power factor of less than unity; and (5) a higher initial cost.

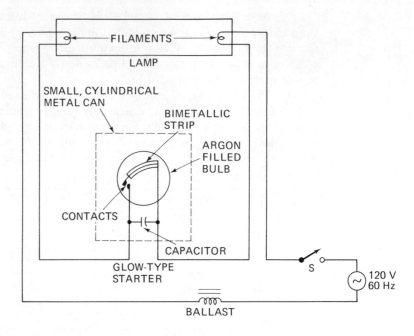

Figure 3-3   Typical fluorescent fixture

The fluorescent lamp is illustrated in Fig. 3-3. A tungsten filament is sealed into each end of a long glass tube whose inner surface is coated with a suitable *phosphor*. Essentially all air remaining in the tube is then pumped out and very small volumes of *argon* gas and mercury are introduced.

In operation, the passage of current through the filaments produces sufficient heat to vaporize the mercury so that the tube is filled with mercury vapor. Then, a relatively high voltage is impressed across the tube, using the filaments as electrodes. The high voltage *ionizes** the argon and mercury vapor and so produces ultraviolet rays. When these rays strike the phosphor coating, the coating *fluoresces* (glows) and produces visible light. Once the tube starts to glow, the filament circuit is opened but the heat produced by ionization of the gases keeps the mercury vaporized so that the lamp continues to emit light.

As shown in Fig. 3-3, the filament circuit is controlled by what is called a *glow-type starter*. Its movable and fixed contact points are enclosed in a small glass bulb containing argon gas. The movable contact point is attached to a *bimetallic strip*.† (Normally the fixed point is not; that is, the two points are

---

*   Recall from Sec. 1-7, that when valence electrons break away from their parent atoms, the atoms become ions. This process is called ionization.

†   A bimetallic strip consists of two strips of dissimilar metal that are welded or riveted together along their length. One end of the strip is rigidly fixed and the other carries a contact. Heat causes the metal strips to expand. One metal expands more than the other so that the arm bends.

separated.) But when the main switch is closed, the 120-volt current ionizes the gas in the bulb so that a small current flows between the movable and fixed contact points in the bulb. This current produces heat which in turn causes the bimetallic strip to bend until the two contact points touch each other. When that occurs, the flow of current through the gas ceases, but it now flows instead through the closed contact points and then through the filaments of the lamps.

When current stops flowing through the gas, the bimetallic strip cools off and the contact points separate to open the filament circuit. Enough residual heat remains, however, to keep the contacts closed long enough for the filaments to vaporize the mercury in the lamp.

When ionization occurs within the tube, the voltage across the tube drops to a very low value. Since the starter is connected in parallel with the tube, the voltage across the starter is at the same low value and is insufficient to cause ionization of the argon gas in the starter bulb. As a result, the contact points remain separated and the filament circuit of the lamp remains open.

A device called a *ballast* is connected in series with the lamp and starter and consists of many turns of fine wire on an iron core. Because of self-induction, a high CEMF (counter electromotive force) is generated by the ballast at the instant that the contact points of the starter separate and break the circuit. This high-voltage surge is applied across the filament electrodes of the lamp, ionizing the mercury vapor. Once this ionization starts, current continues to flow through the lamp even though the voltage across the electrodes drops to a low value.

After ionization starts, the ballast performs a second function. Once started, ionization tends to grow and, unless limited, the resulting excessive current would destroy the lamp. The ballast, however, provides the needed safety action. After it has set up the high-voltage surge that starts ionization, the ballast acts as a *reactor* that keeps current through the lamp within safe limits.

The capacitor across the starter contact points reduces sparking across these points as they are separated. Without the capacitor, the electrical pressure would cause current to arc across the gap formed by the separating contacts. This arcing would ionize the gap and current would continue to flow even after the contact points were a considerable distance apart. Since the electrical pressure is used to charge the capacitor rather than in forming an arc, sparking is reduced.

Although the fluorescent lamp is essentially an ac device, it can be operated on dc. The high-voltage surge set up at the time the contact points open is produced by either type of current. The protective action of the ballast is absent, however, since no CEMF is generated by dc. It is necessary, therefore, to connect a current-limiting resistor in series with the ballast. This resistor causes a power loss, of course, and there is some deterioration in lamp performance.

The schematic diagram of a typical two-lamp fluorescent *fixture* (often called a *luminaire* in engineering publications) is shown in Fig. 3-4. Here, an autotransformer steps up the supply voltage to the level needed for *starting*. The starters (notice symbol in the figure) permit current flow through the lamp filaments. After a few seconds, the heated filament emits electrons, and as the starter opens, a high voltage causes current to flow from the filament through the mercury vapor to the phosphor coating. The lamp current is then limited by reactor characteristic of the ballast.

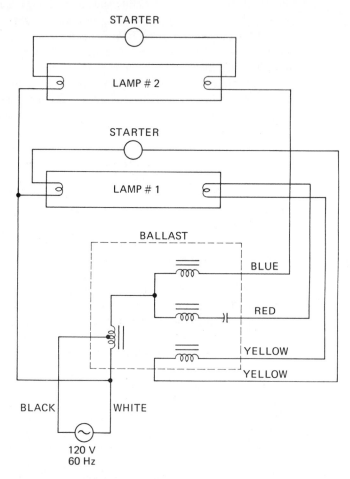

Figure 3-4   Typical two-lamp fluorescent fixture

The capacitor in series with the filament of lamp 1 causes the current to be out of phase with that of lamp 2. As the ac goes through zero when reversing, the lamps then go on and off at different times to avoid a *stroboscopic* effect on moving objects. The capacitor also improves the power factor of the unit to about 0.95. Without the capacitor, the high inductance of the ballasts produces a power factor ranging from 0.5 to 0.75. For large lighting loads, a low power factor is, of course, undesirable.

The type of starting described thus far is called *preheat starting*. Fluorescent lamps designed for preheat starting have bi-pin (two-pin) bases, and are available in sizes from 4 to 100 watts.

A so-called *instant-start* fluorescent lamp also has a filament at each end, but they are short-circuited inside the lamp. Thus, current cannot flow through the filaments before the lamp lights. As we can see in Fig. 3-5, no starter is needed, but the ballast provides an open-circuit voltage of from 450 to 600 V which produces ionization immediately when the switch is closed.

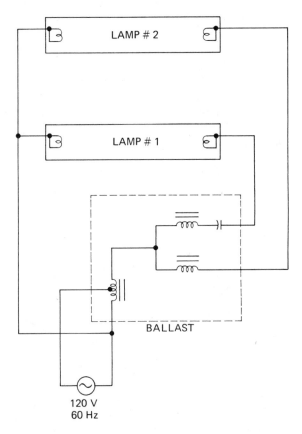

Figure 3-5   Fluorescent fixture using instant-start lamps

Most instant-start lamps are of the *Slimline* ⓣ type and use a single-contact base. Sizes range from 40 to 75 W.

Another form of instant-start lamp is the *cold-cathode* type. Although the internal construction of this lamp is somewhat different than that of the conventional instant-start lamp, its operating principle is essentially the same. The voltages involved are considerably higher, however, being on the order of from 800 to 1,000 volts. For this reason, wire designed for a higher voltage than ordinary fixture wire must be used between the ballast and lamps. The maximum voltage at which ordinary fixture wire can be employed is 300 volts.

Some newer fluorescent lamps are of the *rapid-start type*; that is, starting is delayed only a fraction of a second. No starter is employed and the ballast delivers an open-circuit voltage ranging from about 250 to 400 volts. Non-shorted filaments, located at each end of the tube, are heated during operation of the lamp by special 3-1/2 volt windings on the ballast plus the heat produced by current through the tube. Larger size rapid-start lamps use a recessed double-contact base, while smaller sizes use the bi-pin base. Sizes run from 30 to 220 watts.

A so-called *preheat/rapid-start* lamp is also available. It may be used in a fixture designed for either preheat or rapid-start lamps. This is an exception to the general rule that lamps designed for one type of starting will not fit sockets designed for a different kind.

Fluorescent lamps are manufactured in sizes ranging from six-inches to over four-feet in length. In addition, these lamps are made as straight tubes or in circles and semicircles. Different lamp colors are made by coating the tube

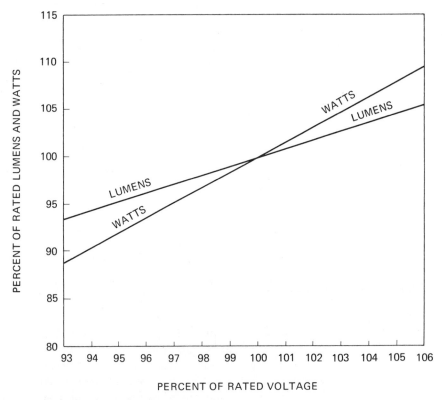

Figure 3-6   Operating characteristics of fluorescent units

interiors with various phosphors. Consequently, fluorescent tubes lend themselves readily to various decorative designs.

Finally, be sure to note that fluorescent lamps, unlike the incandescent types, are *not* rated by voltage. The transformers and ballasts are designed for specific voltages—usually 115, 208, 230, and 265 V.

Performance varies, or course, with deviation from the rated voltages of transformers and ballasts. In Fig. 3-6 the percent of rated lumen output and the changes in power input to ballasts, transformers, and lamps as voltage changes are detailed. Moderate deviations from the rated voltages have little effect on lamp life.

3.4   *How to Obtain Good Lighting*

The preferred method for general lighting calculations is called the Zonal Cavity Method, from the technique of dividing a room into *zones* (also called *cavities*). This method is preferred—not because it is *necessarily* more accurate, but because it is more flexible, can be applied to every type of interior space, and allows more of the actual lighting situation to be taken into account during the calculations. Numerical results are therefore generally more representative of the actual lighting situation than those arrived at by older calculation systems, especially when odd sizes and shapes of rooms are involved, or when surface mounted, recessed, or luminous ceiling systems are used.

Calculating illumination

Interior general lighting calculations are usually used to determine: **(1)** how many luminaires (fixtures) are needed to provide an average given illumination level in an interior space; and **(2)** how the luminaires should be arranged to provide uniform illumination throughout the space. (A luminaire is a complete lighting unit and consists of a light source, together with a globe, reflector, refractor, socket, housing, and other parts integral to the housing.)

Conversely, if the quantity and type of equipment is known for a given space the illumination level may then be calculated.

Illumination calculations are based, of course, on the following equation:

$$fc = \frac{lm}{ft^2} \tag{3-1}$$

It simply expresses the fact that a correct determination of illumination is the result of the lumens (lm) generated, divided by the lighted area in square feet (ft^2).

Typical Luminaires and Luminaire Maintenance Category

1. 2-lamp aluminum industrial with 35° crosswise and lengthwise shielding — LDD Maint. Category II

2. 2-lamp prismatic lens bottom unit with open top — LDD Maint. Category VI

3. Direct-indirect with metal or dense diffusing sides and 35° x 45° louver shielding — LDD Maint. Category II

Coefficients of Utilization for 20 Per Cent Effective Floor Cavity Reflectance, ρ_{FC}

Luminaire 1 — 2-lamp aluminum industrial (Max. S/MH$_{wp}$ = 1.5)

ρ_{CC}	80			70			50			30			10			0
ρ_W	50	30	10	50	30	10	50	30	10	50	30	10	50	30	10	0
RCR																
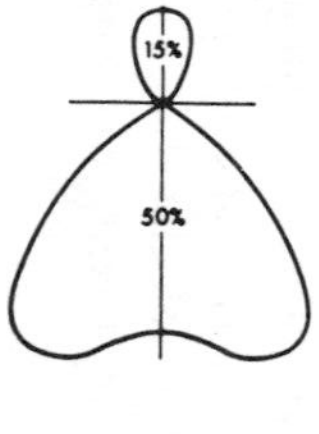 1	.66	.64	.62	.64	.62	.60	.58	.57	.55	.53	.52	.51	.49	.48	.47	.45 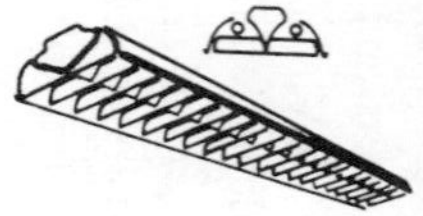
2	.60	.56	.54	.57	.54	.52	.53	.50	.48	.49	.47	.45	.45	.43	.42	.40
3	.54	.50	.46	.52	.48	.45	.48	.45	.42	.44	.42	.40	.41	.39	.37	.36
4	.49	.44	.41	.47	.43	.40	.44	.40	.38	.40	.38	.35	.38	.35	.34	.32
5	.44	.39	.35	.41	.38	.35	.39	.36	.33	.37	.34	.31	.34	.32	.30	.28
6	.40	.35	.31	.38	.34	.31	.36	.32	.29	.33	.30	.28	.31	.29	.27	.25
7	.36	.31	.28	.35	.30	.27	.33	.29	.26	.30	.27	.25	.28	.26	.24	.22
8	.33	.28	.24	.31	.27	.24	.29	.26	.23	.28	.24	.22	.26	.23	.21	.20
9	.29	.25	.21	.28	.24	.21	.27	.23	.20	.25	.21	.19	.23	.20	.18	.17
10	.27	.22	.19	.26	.21	.18	.24	.20	.18	.23	.19	.17	.21	.18	.16	.15

Luminaire 2 — 2-lamp prismatic lens bottom (Max. S/MH$_{wp}$ = 1.2)

ρ_{CC}	80			70			50			30			10			0
ρ_W	50	30	10	50	30	10	50	30	10	50	30	10	50	30	10	0
RCR																
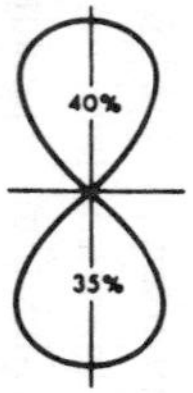 1	.70	.68	.65	.65	.63	.61	.55	.53	.52	.45	.45	.44	.37	.36	.36	.32 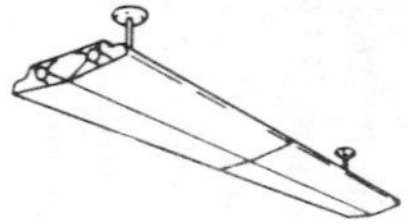
2	.62	.58	.55	.58	.54	.51	.49	.47	.44	.41	.39	.38	.34	.33	.32	.28
3	.56	.50	.47	.52	.47	.44	.44	.41	.38	.37	.35	.33	.30	.29	.28	.25
4	.49	.44	.40	.46	.41	.38	.40	.36	.33	.33	.31	.29	.28	.26	.25	.22
5	.44	.39	.34	.41	.36	.33	.36	.32	.29	.30	.27	.25	.25	.23	.22	.20
6	.40	.34	.30	.37	.32	.29	.32	.28	.25	.27	.24	.22	.23	.21	.19	.17
7	.36	.30	.26	.34	.28	.25	.29	.25	.22	.25	.22	.19	.21	.19	.17	.15
8	.33	.27	.23	.30	.25	.22	.26	.22	.20	.22	.20	.17	.19	.16	.15	.13
9	.29	.24	.20	.27	.22	.19	.24	.20	.17	.20	.17	.15	.17	.15	.13	.11
10	.27	.21	.18	.25	.20	.17	.22	.18	.15	.19	.15	.13	.16	.13	.12	.11

Luminaire 3 — Direct-indirect (Max. S/MH$_{rp}$ = 1.3)

ρ_{CC}	80			70			50			30			10			0
ρ_W	50	30	10	50	30	10	50	30	10	50	30	10	50	30	10	0
RCR																
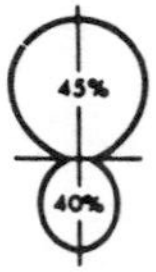 1	.79	.76	.73	.73	.70	.68	.62	.60	.58	.51	.50	.49	.42	.41	.40	.36 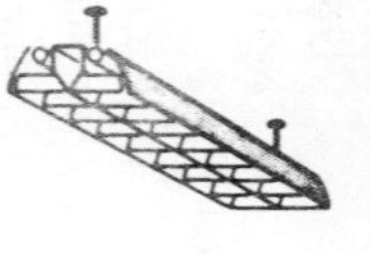
2	.70	.65	.61	.65	.61	.57	.55	.52	.50	.46	.44	.42	.38	.36	.35	.32
3	.62	.56	.52	.58	.53	.49	.49	.46	.43	.41	.39	.36	.34	.33	.31	.28
4	.55	.49	.44	.51	.46	.42	.44	.40	.37	.37	.34	.32	.31	.29	.27	.24
5	.49	.43	.38	.46	.40	.36	.40	.35	.32	.34	.30	.28	.28	.25	.24	.21
6	.44	.38	.33	.41	.35	.31	.36	.31	.28	.30	.27	.24	.25	.23	.21	.19
7	.40	.34	.29	.37	.31	.27	.32	.28	.24	.27	.24	.21	.23	.20	.19	.18
8	.35	.29	.25	.33	.27	.24	.29	.24	.21	.24	.21	.18	.20	.18	.16	.14
9	.32	.25	.22	.30	.24	.21	.26	.21	.18	.22	.19	.16	.19	.16	.14	.12
10	.30	.23	.19	.28	.22	.19	.24	.20	.16	.20	.17	.14	.17	.14	.12	.12

Typical Distribution and Maximum Spacing

- Luminaire 1: 15% / 50% — Max. S/MH$_{wp}$ = 1.5
- Luminaire 2: 40% / 35% — Max. S/MH$_{wp}$ = 1.2
- Luminaire 3: 45% / 40% — Max. S/MH$_{rp}$ = 1.3

Figure 3-7   Coefficients for three fluorescent and three incandescent fixtures (Reprinted with permission from Chapter 9 of Illuminating Engineering Society Handbook, 5th Ed.)

Typical Luminaires and Luminaire Maintenance Category

Typical Distribution and Maximum Spacing[a]

Coefficients of Utilization for 20 Per Cent Effective Floor Cavity Reflectance, p_{FC}

Enclosed reflector with incandescent lamp. LDD Maint. Category V.

Typical Distribution 0% / 70% — Max. $S/MH_{wp} = 1.5$

p_{CC} →	80			70			50			30			10			0
p_W →	50	30	10	50	30	10	50	30	10	50	30	10	50	30	10	0
RCR[b]																
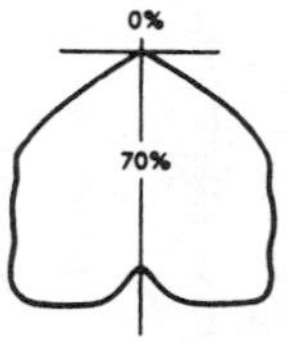 1	.78	.77	.74	.76	.75	.73	.74	.72	.71	.71	.70	.68	.68	.67	.66	.65 
2	.72	.68	.66	.71	.67	.65	.68	.66	.63	.65	.64	.62	.64	.62	.61	.59
3	.66	.62	.59	.65	.61	.58	.63	.60	.57	.61	.59	.56	.60	.57	.55	.54
4	.60	.56	.53	.59	.55	.52	.58	.54	.51	.56	.53	.51	.55	.53	.50	.49
5	.55	.50	.47	.54	.50	.46	.53	.49	.46	.52	.48	.46	.51	.48	.45	.44
6	.51	.45	.42	.50	.45	.42	.49	.45	.41	.48	.44	.41	.47	.43	.41	.40
7	.46	.41	.37	.46	.41	.37	.45	.40	.37	.44	.40	.37	.43	.39	.36	.35
8	.42	.37	.33	.42	.37	.33	.41	.36	.33	.40	.36	.33	.39	.35	.33	.31
9	.39	.33	.30	.38	.33	.30	.37	.33	.30	.37	.32	.29	.36	.32	.29	.28
10	.33	.28	.25	.33	.28	.25	.32	.28	.25	.32	.28	.24	.31	.27	.24	.23

Medium distribution reflector and concave lens. LDD Maint. Category V.

Typical Distribution 0% / 60% — Max. $S/MH_{wp} = 1.3$

p_{CC} →	80			70			50			30			10			0
p_W →	50	30	10	50	30	10	50	30	10	50	30	10	50	30	10	0
RCR[b]																
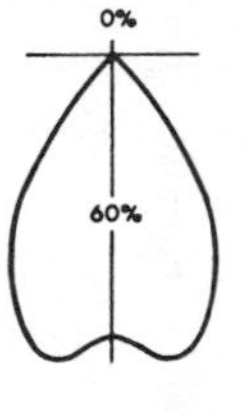 1	.66	.65	.63	.64	.63	.62	.62	.61	.60	.60	.59	.58	.58	.57	.57	.55 
2	.61	.59	.56	.61	.58	.56	.58	.56	.54	.56	.55	.53	.55	.53	.52	.51
3	.57	.54	.51	.56	.53	.51	.55	.52	.50	.53	.51	.49	.52	.50	.48	.47
4	.53	.49	.46	.52	.49	.46	.51	.48	.45	.49	.47	.45	.48	.47	.44	.43
5	.49	.45	.42	.48	.45	.42	.47	.44	.41	.46	.43	.41	.45	.43	.41	.40
6	.45	.42	.39	.45	.41	.39	.44	.41	.38	.43	.40	.38	.42	.40	.38	.37
7	.42	.38	.36	.41	.38	.35	.41	.38	.35	.40	.37	.35	.40	.37	.35	.34
8	.39	.35	.33	.38	.35	.32	.38	.35	.32	.37	.34	.32	.37	.34	.32	.31
9	.36	.32	.30	.35	.32	.30	.35	.32	.29	.35	.32	.29	.34	.31	.29	.28
10	.32	.28	.25	.32	.28	.25	.31	.28	.25	.31	.27	.25	.30	.27	.25	.24

Wide distribution reflector and flat lens. LDD Maint. Category V.

Typical Distribution 0% / 65% — Max. $S/MH_{wp} = 1.2$

p_{CC} →	80			70			50			30			10			0
p_W →	50	30	10	50	30	10	50	30	10	50	30	10	50	30	10	0
RCR[b]																
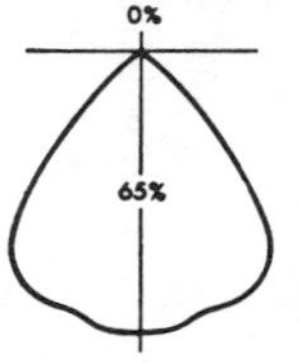 1	.75	.73	.71	.73	.71	.70	.70	.69	.68	.68	.67	.66	.66	.65	.64	.63 
2	.68	.65	.62	.67	.64	.62	.65	.62	.60	.63	.61	.59	.61	.59	.58	.57
3	.63	.59	.55	.62	.58	.54	.60	.57	.53	.58	.56	.52	.57	.55	.52	.51
4	.58	.53	.50	.57	.53	.50	.55	.52	.49	.54	.51	.48	.53	.50	.48	.47
5	.53	.48	.45	.52	.48	.45	.51	.47	.44	.50	.46	.44	.49	.46	.43	.42
6	.48	.44	.41	.48	.44	.41	.47	.44	.40	.46	.43	.40	.45	.42	.40	.39
7	.45	.40	.37	.45	.40	.37	.44	.40	.37	.43	.39	.37	.42	.39	.36	.35
8	.42	.37	.34	.41	.37	.34	.40	.36	.33	.40	.36	.33	.39	.36	.33	.32
9	.38	.33	.30	.38	.33	.30	.37	.33	.30	.36	.32	.30	.36	.32	.30	.28
10	.35	.30	.27	.35	.30	.27	.34	.30	.27	.34	.30	.27	.33	.29	.27	.26

[a] Ratio of maximum spacing between luminaire centers to mounting (or ceiling) height above the work plane. See "Luminaire Spacing"
[b] RCR = Room Cavity Ratio.
[c] p_{CC} = Per cent effective ceiling cavity reflectance.
[d] p_W = Per cent wall reflectance.

Figure 3-7   Continued

But this basic equation is for the ideal, theoretical case. It does not take into account the varying degrees of absorbtion of light by wall, ceiling, and floor surfaces, the interreflection of light, the efficiency or distribution of the luminaire, the shape of the room, or even the point at which the illumination is measured—floor, desk top, work bench, or the like.

Factoring in one additional quantity, however, measures the ideal or theoretical problem both realistically and practically since it takes into account all the influences mentioned above. This additional quantity is called the *Coefficient of Utilization* (CU). The basic equation is then expressed as follows:

$$fc = \frac{lm}{ft^2} \times CU \qquad\qquad (3\text{-}2)$$

Be sure to study Equations 3-1 through 3-6 carefully since they develop the Zonal Cavity Method—the systematic way to determine an accurate CU. To account for the depreciation of illumination over time due to dirt on room and luminaire surfaces and aging of the light sources, two other factors are usually added to the basic equation: **(1)** Lamp Lumen Depreciation (LLD); and **(2)** Luminaire Dirt Depreciation (LDD).

The equation is now complete and represents *average maintained illumination* resulting from a given luminaire in a given room. Thus, we have,

$$fc = \frac{lm}{ft^2} \times CU \times LLD \times LDD \qquad\qquad (3\text{-}3)$$

Now, to determine the number of luminaires needed to provide a given average maintained illumination level, we simply rearrange Equation 3-3 in this form:

$$\text{Number of Luminaires} = \frac{\text{Maintained fc} \times ft^2}{CU \times \dfrac{lm}{lamp} \times \dfrac{lamps}{luminaire} \times LLD \times LDD} \qquad (3\text{-}4)$$

An alternative equation which provides luminaire spacing information can be written:

$$\text{Area/Luminaire} = \frac{\dfrac{Lumens}{lamp} \times \dfrac{lamps}{luminaire} \times CU \times LDD \times LLD}{\text{Maintained Footcandles}} \qquad (3\text{-}5)$$

Then the total number of luminaires required is:

$$\text{Total Luminaires} = \frac{\text{Total Room Area}}{\text{Area/Luminaire}} \qquad\qquad (3\text{-}6)$$

Determine the quantities needed to solve the equation by using the following step-by-step procedure.

1.   Select the lamp and luminaire (typical data for making these selections are shown in Fig. 3-7.)

2.   Decide upon the average maintained illumination required (see data in any table of illumination recommendations.)

3.   Calculate the work-plane area (length times width)

4.   Determine the CU (as described below)

5.   Look up the lumens per lamp (lamp manufacturer's catalog)

6.   Look up the LDD and LLD in the IES Handbook (5th Ed.)

The calculation of CU or the coefficient of utilization involves three steps: (1) determining the cavity ratios; (2) determining the effective reflectances; (3) looking up the CU in the reference tables which use the computations that you made in Steps 1 and 2.

Remember that a room with suspended luminaires and a work plane not on the floor has three zones or cavities: ceiling cavity, room cavity, and floor cavity. While the use of surface mounted or recessed luminaires would effectively eliminate the ceiling cavity, of course, and putting the work plane on the floor would eliminate or set aside consideration of the floor cavity, there is always a room cavity in any consideration of lighting relationships.

Two factors, lamp lumen depreciation (LLD) and luminaire dirt depreciation (LDD) were mentioned briefly in deriving Equation 3-4. Let us take a closer look at these two factors.

Lamp lumen depreciation results from lamp aging. You can determine the amount of LLD by referring to the manufacturers' statistics for the performance of each particular type of lamp. Proper choice of this value will give you the LLD factor which accurately measures the relation between the initial output of a group of new lamps and that of the same group of lamps at the time of its planned replacement. As an example, Table 3-3 lists an average of manufacturers' factors for several frequently used lamps at 70 percent of rated average life—this level being the minimum reached in an installation where burnouts are promptly replaced, whether the planned relamping is to be a group or a random replacement program.

Luminaire dirt depreciation results from the accumulation of dirt on the luminaire, particularly on the bulb or globe or reflector. Categories I to VI classify types of luminaires and are based on similar characteristics of dirt attraction and dirt reflection. See Fig. 3-8. Each luminaire category chart contains curves representing average luminaire depreciation in each of five degrees of freedom from dirt—very clean to very dirty. Table 3-4 provides standards for determining which condition prevails. The appropriate curve is

LAMP LUMEN DEPRECIATION FACTORS

	GE Lamp Type	Watts	Mean Lumen Factor* (%)	LLD** (%)
Incandescent	100A	100	93	90
	150A	150	93	90
	150PAR/SP and FL	150	84	78
	150R/SP and FL	150	89	85
	300M/IF	300	91	87
	300R/SP and FL	300	94	92
	500/IF	500	91	88
	1000/IF	1000	92	89
	1500/IF	1500	84	78
Fluorescent	F40CW Mainlighter®	40	87	83
	F40WW "	40	87	83
	F40CWX "	40	83	73
	F40WWX "	40	83	73
	F40CW/S Staybright ®	40	90	86
	F40WW/S "	40	90	86
	F96T12/CW	50	91	87
	F96T12/WW	50	91	87
	F96T12/CWX	50	87	78
	F48T12/CW/HO	60	87	79
	F96T12/CW/HO	110	87	77
	F48PG17/CW	110	77	67
	F96PG17/CW	215	75	65
	F48T12/CW/1500	110	77	67
	F96T12/CW/1500	215	80	71

* Use the Mean Lumen Factor when calculations involve **average** illumination between relampings (based upon lamp light output at 40-50% of rated life depending upon lamp type).

** Use the LLD Factor when calculations involve **minimum** illumination between relampings (based upon lamp light output at approximately 70% of rated life).

† Factors for mercury lamps are based upon 24,000 hours life.

Note: Factors shown for fluorescent lamps are based on 3 hours/start. Factors for Multi-Vapor and Lucalox lamps are based on 10 hours/start. All HID lamp factors are for vertical burning.

Table 3-3   Lamp lumen depreciation factor (reprinted by permission from General Electric Bulletin TPC-42, Nov. 1973)

then used to determine the LDD factor which measures the relation between the light output of the clean luminaire and that of the same luminaire at the time of the next planned cleaning.

To determine luminaire dirt depreciation, select the category curve in Fig. 3-8 which most nearly illustrates a given luminaire's characteristics of dirt attraction or of dirt retention. See also the data in Fig. 3-7.

Next, from Table 3-4 determine the atmosphere in which the luminaire will operate. Then on the appropriate curve in Fig. 3-8, follow the applicable dirt condition curve to the proper vertical line for elapsed time in months of the planned cleaning cycle. Read the LDD on the vertical axis.

Now, let us discuss luminaire spacing; then we can go back and put all

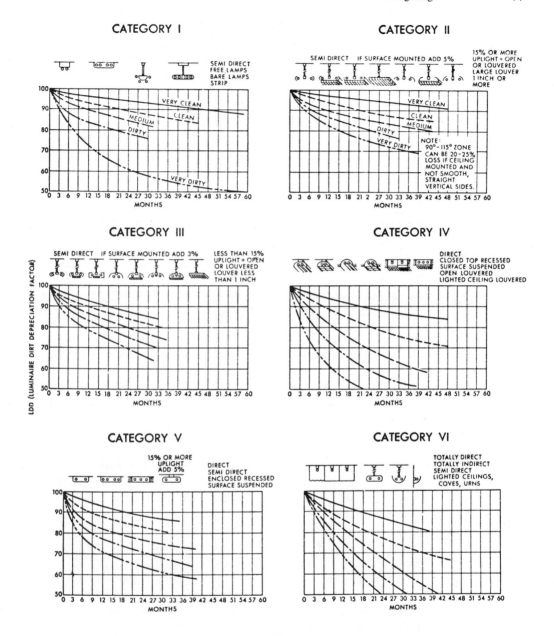

Figure 3-8   The LDD factors for six categories and five degrees of dirtiness (Reprinted with permission from Chapter 9 of Illuminating Engineering Society Handbook, 5th Ed.)

of this information to good use by showing a typical application problem and its solution.

	Very Clean	Clean	Medium	Dirty	Very Dirty
Generated Dirt	None	Very little	Noticeable but not heavy	Accumulates rapidly	Constant accumulation
Ambient Dirt	None (or none enters area)	Some (almost none enters)	Some enters area	Large amount enters area	Almost none excluded
Removal or Filtration	Excellent	Better than average	Poorer than average	Only fans or blowers if any	None
Adhesion	None	Slight	Enough to be visible after some months	High—probably due to oil, humidity, or static	High
Examples	High grade offices, not near production; laboratories; clean rooms	Offices in older buildings or near production; light assembly; inspection	Mill offices; paper processing; light machining	Heat treating; high speed printing; rubber processing	Similar to **Dirty** but luminaires within immediate area of contamination

Table 3-4   Five degrees of dirt conditions (reprinted by permission from Ch. 9 Illuminating Engineering Society Handbook, 5th Ed.)

Because a visual task (the thing to be seen) is likely to be located (or relocated) at any point in the room being lighted, the illumination should remain reasonably uniform at any point in the room. Perfect illumination uniformity is generally not feasible, but acceptable uniformity is considered as achieved when the maximum and minimum values are not more than one-sixth above or below the calculated average.

To develop acceptable uniformity, do not space luminaires too far apart from each other or too far from the walls. Spacing limitations between direct, semi-direct, and general-diffuse luminaires are related to the mounting height above the work plane. For semi-indirect and indirect luminaires, the ceiling height above the work plane is the dimension of reference. Recommended spacing-to-mounting-height-above-the-work-plane ratios, S/MH_{wp}, are given directly below the distribution curve for each luminaire type in Fig. 3-7. These spacing ratios apply to the spacings shown in Fig. 3-9.

The common practice of letting the distance from the luminaires to the wall equal one-half the distance between rows [Fig. 3-10(a)] results in inadequate illumination near the walls. Since, for example, desks and benches are frequently located along the walls, a distance of 2-1/2 feet from the wall to the center of the luminaire should be employed to avoid excessive drop-off in illumination. This locates the luminaires over the edge of desks facing the wall or over the center of desks perpendicular to the wall. See Fig. 3-10(b). To further improve illumination uniformity across the room, it is often desirable to use somewhat closer spacings between outer rows of luminaires than between central rows. Take care that no spacing exceeds the maximum permissible spacing.

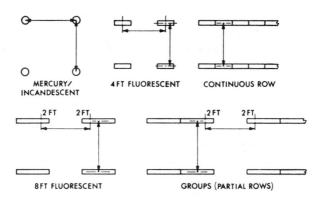

Figure 3-9   Spacing dimensions to be used in relation to spacing-to-mounting height ratio. Mounting height is from luminaires to work plane for direct, semi-direct, and general-diffuse luminaires and from ceiling to work plane for semi-indirect and indirect luminaires (Reprinted with permission from Chapter 9 of Illuminating Engineering Society Handbook, 5th Ed.)

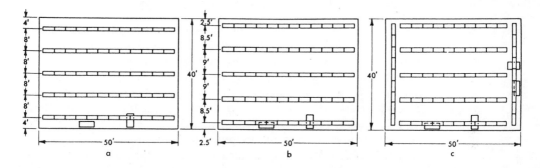

Figure 3-10   Lighting layout using equal spacing between continuous rows of luminaires (a). Layout is changed to provide more illumination near side wall (b). By adding four more four-foot units at each end, layout (b) can be modified to provide 80 percent more light near the end walls and prevent possible scallop effects (c) (Reprinted by permission from Chapter 9 of Illuminating Engineering Society Handbook, 5th Ed.)

To prevent excessive reduction in illumination at the ends of the room it is better to locate the ends of fluorescent luminaire rows 6 to 12 inches from the walls—and in any case no more than 2 feet from the walls. Even the 6- to 12-inch spacing leaves much to be desired from the standpoint of uniformity. Where practicable, use the arrangement shown in Fig. 3-10(c). It is much more satisfactory since, with this arrangement, the units at each end of the row are replaced by a continuous row parallel to and 2-1/2 feet from the end wall. In the example shown, 5 units are replaced by 9 units to provide a potential increase in the illumination at the end of the room of 80 percent over that obtainable with the layout shown in Fig. 3-10(b). This technique not only improves uniformity but also eliminates scallops of light on the end walls. It also provides a uniform wash of light on all four walls.

Another excellent method of compensating for the normal reduction in illumination that may be expected at the ends of rows is to use a greater number of lamps in the end units. Still another technique is to provide additional units between the rows at each end. You can place the units either parallel to or at right angles to the rows.

Spacings closer than the maximum permissible are often highly desirable if you want to reduce harsh shadows and veiling reflections in the work task—as well as to further improve uniformity. This is particularly true for direct and semi-direct equipment. Spacings that are substantially less than that for the maximum permissible spacing should be seriously considered when you install such equipment.

Problem:

The first floor of an old home has been made into a single large room that is to be used as office space by a jewelry distributor. The dimensions are $45'0'' \times 30'0'' \times 10'6''$ and the luminaire mounting height is $9'0''$ and the work plane height is $3'0''$ and reflectances are 80%, 50%, and 30%, (ceiling, walls, and floor, respectively); the illumination level is 150 fc. The luminaire to be used is a 2-lamp prismatic lens, bottom unit with open top. See top of Fig. 3-7. The lamps are GE type F48T12/CW/HO (Table 3-3) and provide approximately 80 lumens per watt. The lights are to be replaced at burnout. From Table 3-3, use an LLD of 79 percent. How many luminaires are needed and what is the most satisfactory arrangement? For convenience, a CU work sheet is illustrated in Fig. 3-11.

Solution:

1.   Fill in the information required at the top, and in Step 1, of Fig. 3-11.

2.   From Table 3-5, the cavity ratios are 0.4, 1.7, and 0.8 for ceiling, room, and floor, respectively.

3.   From Table 3-6, reflectances are 74 and 27, respectively for the effective ceiling and floor cavities.

4.   From Fig. 3-10, the CU is 0.54. Since the effective floor-cavity reflectance is greater than 20 percent, use the multiplication factor of 1.05 indicated in Table 3-4. Then, CU = 0.58

COEFFICIENT OF UTILIZATION WORK SHEET

Job Identification: Office

Name: Empire Jewelry Date: 10-24

Average maintained footcandles: 150

Luminaire Data:	Lamp Data:
Mfr: G.E.	Type: F96/T12/CW/HO
Catalog No: 35	Lumens: 4800
Lamps/Luminaire: 2	LLD: 0.90
LDD: 0.90	Lumens/Luminaire:

Step 1. Fill in sketch

_____ $=h_{CC}$

_____ $=h_{RC}$

_____ $=h_{FC}$

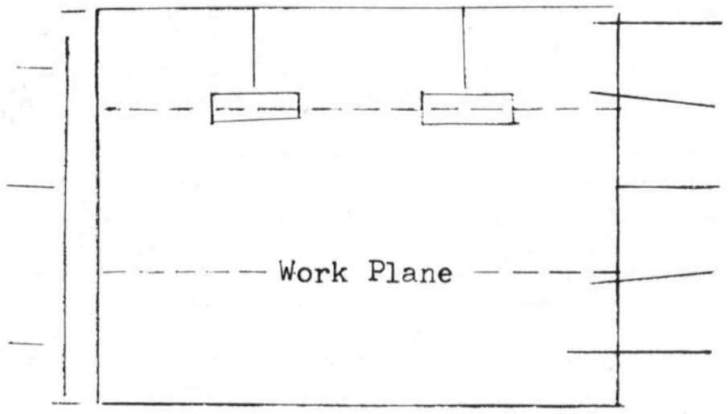

C = 70

W = 50

W = 50

W = 30

F = 30

Length: 45 Width: 30 Height: 10.5

Step 2. Determine cavity ratios from Table 3-5	Ceiling Cavity Ratio	0.4
	Room Cavity Ratio	1.7
	Floor Cavity Ratio	0.8

Step 2. Obtain Effective Ceiling Cavity Reflectance from Table 3-6	74
Obtain Effective Floor Cavity Reflectance from Table 3-6	27
Obtain Coefficient of Utilization (CU) from Fig. 3-7	.54

If the effective floor-cavity reflectance is other than 20%, adjust CU by factor from Table 3-7

Adjusted CU = 0.58

Figure 3-11 CU work sheet

Now, to determine the number of luminaires, you will use Equation 3-4. First, however, you must determine the value of LDD. From Fig. 3-7, you learn that the LDD maintance category is VI. (The atmosphere would be considered clean for this remodeled room.) Then, from Fig. 3-8, LDD = 0.90.

Room Dimensions		Cavity Depth																			
Width	Length	1.0	1.5	2.0	2.5	3.0	3.5	4.0	5.0	6.0	7.0	8	9	10	11	12	14	16	20	25	30
8	8	1.2	1.9	2.5	3.1	3.7	4.4	5.0	6.2	7.5	8.8	10.0	11.2	12.5	—	—	—	—	—	—	—
	10	1.1	1.7	2.2	2.8	3.4	3.9	4.5	5.6	6.7	7.9	9.0	10.1	11.3	12.4	—	—	—	—	—	—
	14	1.0	1.5	2.0	2.5	2.9	3.4	3.9	4.9	5.9	6.9	7.9	8.8	9.7	10.8	11.7	12.2	11.8	—	—	—
	20	0.9	1.3	1.6	2.2	2.6	3.1	3.5	4.4	5.2	6.1	7.0	7.9	8.8	9.6	10.5	11.0	10.0	—	—	—
	40	0.7	1.1	1.5	1.9	2.3	2.6	3.0	3.7	4.5	5.3	6.9	6.5	7.4	8.1	8.8	10.3	11.8	11.7	—	—
10	10	1.0	1.5	2.0	2.5	3.0	3.5	4.0	5.0	6.0	7.0	8.0	9.0	10.0	11.0	12.0	12.0	12.0	—	—	—
	14	0.9	1.3	1.7	2.1	2.6	3.0	3.4	4.3	5.1	6.0	6.9	7.7	8.6	9.5	10.3	10.5	10.6	12.5	—	—
	20	0.7	1.1	1.5	1.9	2.3	2.6	3.0	3.7	4.5	5.3	6.0	6.8	7.5	8.3	9.0	9.4	10.0	11.0	—	—
	30	0.7	1.0	1.3	1.7	2.0	2.3	2.7	3.3	4.0	4.7	5.3	6.0	6.6	7.3	8.0	8.7	8.2	10.2	—	—
	40	0.6	0.9	1.2	1.5	1.9	2.2	2.5	3.1	3.7	4.4	5.0	5.6	6.2	6.9	7.5	7.6	7.8	9.7	—	—
	60	0.6	0.9	1.2	1.5	1.7	2.0	2.3	2.9	3.5	4.1	4.7	5.3	5.9	6.5	7.1	8.2	9.4	11.7	—	—
12	12	0.8	1.2	1.7	2.1	2.5	2.9	3.3	4.2	5.0	5.8	6.7	7.5	8.4	9.2	10.2	11.7	11.6	—	—	—
	16	0.7	1.1	1.5	1.8	2.2	2.6	2.9	3.6	4.4	5.1	5.8	6.6	7.3	8.0	8.7	10.2	11.7	12.5	—	—
	24	0.6	0.9	1.2	1.6	1.9	2.2	2.5	3.1	3.8	4.4	5.0	5.6	6.2	6.9	7.5	8.7	10.0	12.3	—	—
	36	0.6	0.8	1.1	1.3	1.7	2.0	2.1	2.6	3.1	3.6	4.1	4.6	5.1	5.8	6.2	7.2	8.2	10.2	—	—
	50	0.5	0.7	1.0	1.2	1.5	1.7	2.0	2.4	2.9	3.4	3.9	4.4	4.9	5.4	5.8	6.8	7.8	9.8	12.2	—
	70	0.5	0.7	1.0	1.2	1.5	1.7	2.0	2.4	2.9	3.4	3.9	4.4	4.9	5.4	5.9	6.8	7.8	9.7	12.2	—
14	14	0.7	1.1	1.4	1.8	2.1	2.5	2.9	3.6	4.3	5.0	5.7	6.4	7.1	7.8	8.5	9.8	11.4	11.7	—	—
	20	0.6	0.9	1.2	1.5	1.8	2.1	2.4	3.0	3.6	4.2	4.9	5.5	6.1	6.7	7.3	8.7	9.8	12.3	—	—
	30	0.5	0.8	1.0	1.3	1.6	1.8	2.1	2.6	3.1	3.7	4.2	4.7	5.2	5.8	6.3	7.3	8.4	10.5	—	—
	42	0.5	0.7	1.0	1.2	1.4	1.7	1.9	2.4	2.9	3.3	3.8	4.3	4.8	5.2	5.7	6.1	7.0	8.8	11.9	—
	60	0.4	0.7	0.9	1.1	1.3	1.5	1.8	2.2	2.6	3.1	3.5	3.9	4.4	4.8	5.1	6.1	7.0	7.7	11.0	—
	90	0.4	0.6	0.8	1.0	1.2	1.4	1.6	2.0	2.5	2.9	3.3	3.7	4.1	4.5	5.0	5.8	6.6	6.7	10.3	12.4
17	17	0.6	0.9	1.2	1.5	1.8	2.1	2.4	2.9	3.5	4.1	4.7	5.3	5.9	6.5	7.0	8.2	11.4	12.3	12.3	12.4
	25	0.5	0.7	1.0	1.2	1.5	1.7	2.0	2.5	3.0	3.5	4.0	4.5	5.2	5.5	6.0	8.7	10.0	10.5	10.9	11.9
	35	0.4	0.7	0.9	1.1	1.3	1.5	1.7	2.2	2.6	3.1	3.5	3.9	4.3	4.8	5.2	7.3	8.7	9.5	9.7	10.1
	50	0.4	0.6	0.8	1.0	1.2	1.4	1.6	2.0	2.4	2.8	3.1	3.5	3.9	4.3	4.7	6.1	7.6	8.8	9.0	8.4
	80	0.3	0.5	0.7	0.8	1.0	1.2	1.3	1.7	2.0	2.3	2.7	3.0	3.4	3.7	4.0	5.8	7.0	7.7	8.4	10.1
	120	0.3	0.5	0.7	0.8	1.0	1.2	1.3	1.7	2.0	2.3	2.6	3.0	3.4	3.7	4.0	4.7	6.6	6.7	8.3	10.1
20	20	0.5	0.7	1.0	1.2	1.5	1.7	2.0	2.5	3.0	3.5	4.0	4.5	5.1	5.5	6.0	8.2	9.4	10.0	12.5	12.4
	30	0.4	0.6	0.8	1.0	1.2	1.5	1.7	2.1	2.5	2.9	3.3	3.7	4.1	4.5	4.9	6.1	8.2	8.2	10.3	10.9
	45	0.4	0.5	0.7	0.9	1.1	1.3	1.4	1.8	2.2	2.5	2.9	3.2	3.6	4.0	4.3	5.7	6.4	7.2	9.1	10.1
	60	0.3	0.5	0.7	0.8	1.0	1.2	1.3	1.7	2.0	2.3	2.7	3.0	3.4	3.7	4.0	5.1	5.8	6.7	7.5	9.0
	90	0.3	0.5	0.6	0.8	0.9	1.1	1.2	1.5	1.8	2.1	2.4	2.7	2.9	3.3	3.4	4.0	4.6	5.7	6.5	8.6
	150	0.3	0.4	0.6	0.7	0.8	1.0	1.1	1.4	1.7	2.0	2.3	2.6	3.4	3.2	3.4	4.7	4.6	5.7	7.2	8.6
24	24	0.4	0.6	0.8	1.0	1.2	1.5	1.7	2.1	2.5	2.9	3.3	3.7	4.1	4.5	5.0	5.8	8.0	8.2	10.3	12.4
	32	0.4	0.5	0.7	0.9	1.1	1.3	1.5	1.8	2.2	2.6	2.9	3.3	3.6	4.0	4.3	5.1	6.8	7.2	9.0	11.0
	50	0.3	0.5	0.6	0.7	0.9	1.1	1.2	1.5	1.8	2.2	2.5	2.8	3.1	3.4	3.7	4.4	5.4	6.2	7.8	9.4
	70	0.3	0.4	0.6	0.7	0.8	1.0	1.1	1.4	1.7	2.0	2.2	2.5	2.8	3.0	3.3	3.7	4.8	5.5	6.5	8.2
	100	0.3	0.4	0.5	0.6	0.8	0.9	1.0	1.3	1.6	1.8	2.1	2.4	2.6	2.9	3.1	3.7	4.2	5.2	6.5	7.9
	160	0.2	0.4	0.5	0.6	0.7	0.8	1.0	1.2	1.4	1.7	1.9	2.1	2.4	2.6	2.8	3.3	3.8	4.7	5.9	7.1

Table 3-5   Cavity ratios (reprinted by permission from Ch. 9 IES Handbook, 5th Ed.)

Room Dimensions		Cavity Depth																			
Width	Length	1.0	1.5	2.0	2.5	3.0	3.5	4.0	5.0	6.0	7.0	8	9	10	11	12	14	16	20	25	30
30	30	0.3	0.5	0.7	0.8	1.0	1.2	1.3	1.7	2.0	2.3	2.7	3.0	3.3	3.7	4.0	4.7	5.4	6.7	8.4	10.0
	45	0.3	0.4	0.6	0.7	0.8	1.0	1.1	1.4	1.7	1.9	2.2	2.5	2.7	3.0	3.3	3.8	4.4	5.5	6.9	8.2
	60	0.3	0.4	0.5	0.6	0.7	0.9	1.0	1.2	1.5	1.7	2.0	2.2	2.5	2.7	3.0	3.5	4.0	5.0	6.2	7.4
	90	0.2	0.3	0.4	0.6	0.7	0.8	0.9	1.1	1.3	1.6	1.8	2.0	2.2	2.5	2.7	3.1	3.6	4.5	5.6	6.7
	150	0.2	0.3	0.4	0.5	0.6	0.7	0.8	1.0	1.2	1.4	1.6	1.8	2.0	2.2	2.4	2.8	3.2	4.0	5.0	5.9
	200	0.2	0.3	0.4	0.5	0.6	0.7	0.8	1.0	1.1	1.3	1.5	1.7	1.9	2.0	2.2	2.6	3.0	3.7	4.7	5.6
36	36	0.3	0.4	0.6	0.7	0.8	1.0	1.1	1.4	1.7	1.9	2.2	2.5	2.8	3.0	3.3	3.9	4.4	5.5	6.9	8.3
	50	0.2	0.3	0.5	0.6	0.7	0.8	1.0	1.2	1.4	1.7	1.9	2.1	2.5	2.6	2.9	3.3	3.8	4.8	5.9	7.2
	75	0.2	0.3	0.4	0.5	0.6	0.7	0.8	1.0	1.2	1.4	1.6	1.8	2.0	2.3	2.5	2.9	3.3	4.1	5.1	6.1
	100	0.2	0.3	0.4	0.5	0.6	0.7	0.8	0.9	1.1	1.3	1.5	1.7	1.9	2.1	2.3	2.6	3.0	3.8	4.9	5.7
	150	0.2	0.2	0.3	0.4	0.5	0.6	0.7	0.8	1.0	1.2	1.3	1.6	1.7	1.9	2.1	2.4	2.8	3.5	4.3	5.2
	200	0.2	0.2	0.3	0.4	0.5	0.6	0.7	0.8	1.0	1.1	1.3	1.5	1.6	1.8	2.0	2.3	2.6	3.3	4.1	4.9
42	42	0.2	0.4	0.5	0.6	0.7	0.8	1.0	1.2	1.4	1.6	1.9	2.1	2.4	2.6	2.8	3.3	3.8	4.7	5.9	7.1
	60	0.2	0.3	0.4	0.5	0.6	0.7	0.8	1.0	1.2	1.4	1.6	1.8	2.0	2.2	2.4	2.8	3.2	4.0	5.0	6.0
	90	0.2	0.3	0.3	0.4	0.5	0.6	0.7	0.9	1.0	1.2	1.4	1.6	1.7	1.9	2.1	2.4	2.8	3.5	4.4	5.2
	140	0.1	0.2	0.3	0.4	0.5	0.5	0.6	0.8	0.9	1.1	1.2	1.4	1.5	1.7	1.9	2.2	2.5	3.1	3.9	4.4
	200	0.1	0.2	0.3	0.3	0.4	0.5	0.6	0.7	0.9	1.0	1.1	1.3	1.4	1.6	1.7	2.0	2.3	3.0	3.6	4.3
	300	0.1	0.2	0.2	0.3	0.4	0.5	0.5	0.7	0.8	0.9	1.1	1.3	1.4	1.5	1.7	1.9	2.2	2.8	3.5	4.2
50	50	0.2	0.3	0.4	0.5	0.6	0.7	0.8	1.0	1.2	1.4	1.6	1.8	2.0	2.2	2.4	2.8	3.2	4.0	5.0	6.0
	70	0.2	0.2	0.3	0.4	0.5	0.6	0.7	0.9	1.0	1.2	1.4	1.5	1.7	1.9	2.0	2.4	2.7	3.4	4.3	5.1
	100	0.1	0.2	0.3	0.4	0.5	0.5	0.6	0.7	0.9	1.0	1.2	1.3	1.5	1.6	1.8	2.1	2.4	3.0	3.7	4.5
	150	0.1	0.2	0.3	0.3	0.4	0.5	0.5	0.7	0.8	0.9	1.1	1.2	1.3	1.5	1.6	1.9	2.1	2.7	3.3	4.0
	300	0.1	0.2	0.2	0.3	0.3	0.4	0.5	0.6	0.7	0.8	0.9	1.0	1.1	1.3	1.4	1.6	1.9	2.3	2.9	3.6
60	60	0.2	0.2	0.3	0.4	0.5	0.6	0.7	0.8	1.0	1.2	1.3	1.5	1.7	1.8	2.0	2.3	2.7	3.3	4.2	5.0
	100	0.1	0.2	0.3	0.3	0.4	0.5	0.5	0.7	0.8	0.9	1.1	1.2	1.3	1.5	1.6	1.9	2.1	2.7	3.3	4.0
	150	0.1	0.2	0.2	0.3	0.4	0.4	0.5	0.6	0.7	0.8	0.9	1.0	1.2	1.3	1.4	1.6	1.9	2.3	2.9	3.5
	300	0.1	0.1	0.2	0.2	0.3	0.3	0.4	0.5	0.6	0.7	0.8	0.9	1.0	1.1	1.2	1.4	1.6	2.1	2.5	3.0
75	75	0.2	0.2	0.3	0.4	0.4	0.5	0.5	0.7	0.8	0.9	1.1	1.2	1.3	1.5	1.6	1.9	2.1	2.7	3.3	4.0
	120	0.1	0.2	0.2	0.3	0.3	0.4	0.4	0.5	0.6	0.8	0.9	1.0	1.1	1.2	1.3	1.5	1.7	2.2	2.7	3.3
	200	0.1	0.1	0.2	0.2	0.3	0.3	0.4	0.5	0.5	0.6	0.7	0.8	0.9	1.0	1.1	1.3	1.5	1.8	2.3	2.7
	300	0.1	0.1	0.2	0.2	0.2	0.3	0.3	0.4	0.5	0.6	0.7	0.7	0.8	0.9	1.0	1.2	1.3	1.7	2.1	2.5
100	100	0.2	0.2	0.3	0.3	0.3	0.3	0.4	0.5	0.6	0.7	0.8	0.9	1.0	1.1	1.2	1.4	1.6	2.0	2.5	3.0
	200	0.1	0.1	0.2	0.2	0.2	0.3	0.3	0.4	0.4	0.5	0.5	0.7	0.7	0.8	0.9	1.0	1.2	1.5	1.9	2.2
	300	0.1	0.1	0.2	0.2	0.2	0.2	0.3	0.3	0.4	0.5	0.5	0.6	0.7	0.7	0.8	0.9	1.1	1.3	1.7	2.0
150	150	0.1	0.1	0.1	0.1	0.1	0.2	0.3	0.3	0.4	0.5	0.5	0.6	0.7	0.7	0.8	0.9	1.1	1.3	1.7	2.0
	300	—	0.1	0.1	0.1	0.1	0.1	0.2	0.2	0.3	0.3	0.4	0.5	0.5	0.6	0.6	0.7	0.8	1.0	1.2	1.5
200	200	0.1	0.1	0.1	0.1	0.1	0.2	0.2	0.2	0.3	0.3	0.4	0.5	0.5	0.6	0.6	0.7	0.8	1.0	1.2	1.5
	300	0.1	0.1	0.1	0.1	0.1	0.1	0.2	0.2	0.2	0.3	0.3	0.4	0.4	0.5	0.5	0.6	0.7	0.8	1.0	1.2
300	300	—	—	—	—	0.1	0.1	0.1	0.2	0.2	0.2	0.3	0.3	0.3	0.4	0.4	0.5	0.5	0.6	0.7	0.8
500	500	—	—	—	—	0.1	0.1	0.1	0.1	0.1	0.1	0.2	0.2	0.2	0.2	0.2	0.3	0.3	0.4	0.5	0.6

Table 3-5   Continued

Per Cent Ceiling or Floor Reflectance	90				80				70			50			30				10		
Per Cent Wall Reflectance	90	70	50	30	80	70	50	30	70	50	30	70	50	30	65	50	30	10	50	30	10
0	90	90	90	90	80	80	80	80	70	70	70	50	50	50	30	30	30	30	10	10	10
0.1	90	89	88	87	79	79	78	78	69	69	68	59	49	48	30	30	29	29	10	10	10
0.2	89	88	86	85	79	78	77	76	68	67	66	49	48	47	30	29	29	28	10	10	9
0.3	89	87	85	83	78	77	75	74	68	66	64	49	47	46	30	29	28	27	10	10	9
0.4	88	86	83	81	78	76	74	72	67	65	63	48	46	45	30	29	27	26	11	10	9
0.5	88	85	81	78	77	75	73	70	66	64	61	48	46	44	29	28	27	25	11	10	9
0.6	88	84	80	76	77	75	71	68	65	62	59	47	45	43	29	28	26	25	11	10	9
0.7	88	83	78	74	76	74	70	66	65	61	58	47	44	42	29	28	26	24	11	10	8
0.8	87	82	77	73	75	73	69	65	64	60	56	47	43	41	29	27	25	23	11	10	8
0.9	87	81	76	71	75	72	68	63	63	59	55	46	43	40	29	27	25	22	11	9	8
1.0	86	80	74	69	74	71	66	61	63	58	53	46	42	39	29	27	24	22	11	9	8
1.1	86	79	73	67	74	71	65	60	62	57	52	46	41	38	29	26	24	21	11	9	8
1.2	86	78	72	65	73	70	64	58	61	56	50	45	41	37	29	26	23	20	12	9	7
1.3	85	78	70	64	73	69	63	57	61	55	49	45	40	36	29	26	23	20	12	9	7
1.4	85	77	69	62	72	68	62	55	60	54	48	45	40	35	28	26	22	19	12	9	7
1.5	85	76	68	61	72	68	61	54	59	53	47	44	39	34	28	25	22	18	12	9	7
1.6	85	75	66	59	71	67	60	53	59	52	45	44	39	33	28	25	21	18	12	9	7
1.7	84	74	65	58	71	66	59	52	58	51	44	44	38	32	28	25	21	17	12	9	7
1.8	84	73	64	56	70	65	58	50	57	50	43	43	37	32	28	25	21	17	12	9	6
1.9	84	73	63	55	70	65	57	49	57	49	42	43	37	31	28	25	20	16	12	9	6
2.0	83	72	62	53	69	64	56	48	56	48	41	43	37	30	28	24	20	16	12	9	6
2.1	83	71	61	52	69	63	55	47	56	47	40	43	36	29	28	24	20	16	13	9	6
2.2	83	70	60	51	68	63	54	45	55	46	39	42	36	29	28	24	19	15	13	9	6
2.3	83	69	59	50	68	62	53	44	54	46	38	42	35	28	28	24	19	15	13	9	6
2.4	82	68	58	48	67	61	52	43	54	45	37	42	35	27	28	24	19	14	13	9	6
2.5	82	68	57	47	67	61	51	42	53	44	36	41	34	27	27	23	18	14	13	9	6
2.6	82	67	56	46	66	60	50	41	53	43	35	41	34	26	27	23	18	13	13	9	5
2.7	82	66	55	45	66	60	49	40	52	43	34	41	33	26	27	23	18	13	13	9	5
2.8	81	66	54	44	66	59	48	39	52	42	33	41	33	25	27	23	18	13	13	9	5
2.9	81	65	53	43	65	58	48	38	51	41	33	40	33	25	27	23	17	12	13	9	5
3.0	81	64	52	42	65	58	47	38	51	40	32	40	32	24	27	22	17	12	13	8	5
3.1	80	64	51	41	64	57	46	37	50	40	31	40	32	24	27	22	17	12	13	8	5
3.2	80	63	50	40	64	57	45	36	50	39	30	40	31	23	27	22	16	11	13	8	5
3.3	80	62	49	39	64	56	44	35	49	39	30	39	31	23	27	22	16	11	13	8	5
3.4	80	62	48	38	63	56	44	34	49	38	29	39	31	22	27	22	16	11	13	8	5
3.5	79	61	48	37	63	55	43	33	48	38	29	39	30	22	26	22	16	11	13	8	5
3.6	79	60	47	36	62	54	42	33	48	37	28	39	30	21	26	21	15	10	13	8	5
3.7	79	60	46	35	62	54	42	32	48	37	27	38	30	21	26	21	15	10	13	8	4
3.8	79	59	45	35	62	53	41	31	47	36	27	38	29	21	26	21	15	10	13	8	4
3.9	78	59	45	34	61	53	40	30	47	36	26	38	29	20	26	21	15	10	13	8	4
4.0	78	58	44	33	61	52	40	30	46	35	26	38	29	20	26	21	15	9	13	8	4
4.1	78	57	43	32	60	52	39	29	46	35	25	37	28	20	26	21	14	9	13	8	4
4.2	78	57	43	32	60	51	39	29	46	34	25	37	28	19	26	20	14	9	13	8	4
4.3	78	56	42	31	60	51	38	28	45	34	25	37	28	19	26	20	14	9	13	8	4
4.4	77	56	41	30	59	51	38	28	45	34	24	37	27	19	26	20	14	8	13	8	4
4.5	77	55	41	30	59	50	37	27	45	33	24	37	27	19	25	20	14	8	14	8	4
4.6	77	55	40	29	59	50	37	26	44	33	24	36	27	18	25	20	14	8	14	8	4
4.7	77	54	40	29	58	49	36	26	44	33	23	36	26	18	25	20	13	8	14	8	4
4.8	76	54	39	28	58	49	36	25	44	32	23	36	26	18	25	19	13	8	14	8	4
4.9	76	53	38	28	58	49	35	25	44	32	23	36	26	18	25	19	13	7	14	8	4
5.0	76	53	38	27	57	48	35	25	43	32	22	36	26	17	25	19	13	7	14	8	4

Ceiling or Floor Cavity Ratio

Table 3-6   Percent effective ceiling or floor cavity reflectance for various reflectance combinations (reprinted by permission from Ch. 9 IES Handbook, 5th Ed.). For cavity dimensions other than those shown here, use the formula

$$CR = \frac{5 \times h_{CC}(L + W)}{L \times W}$$

For 30 per cent effective floor cavity reflectance, multiply by appropriate factor below.
For 10 per cent effective floor cavity reflectance, divide by appropriate factor below.

Per Cent Effective Ceiling Cavity Reflectance, ρ_{CC}	80			70			50			10		
Per Cent Wall Reflectance, ρ_W	50	30	10	50	30	10	50	30	10	50	30	10
Room Cavity Ratio												
1	1.08	1.08	1.07	1.07	1.06	1.06	1.05	1.04	1.04	1.01	1.01	1.01
2	1.07	1.06	1.05	1.06	1.05	1.04	1.04	1.03	1.03	1.01	1.01	1.01
3	1.05	1.04	1.03	1.05	1.04	1.03	1.03	1.03	1.02	1.01	1.01	1.01
4	1.05	1.03	1.02	1.04	1.03	1.02	1.03	1.02	1.02	1.01	1.01	1.00
5	1.04	1.03	1.02	1.03	1.02	1.02	1.02	1.02	1.01	1.01	1.01	1.00
6	1.03	1.02	1.01	1.03	1.02	1.01	1.02	1.02	1.01	1.01	1.01	1.00
7	1.03	1.02	1.01	1.03	1.02	1.01	1.02	1.01	1.01	1.01	1.01	1.00
8	1.03	1.02	1.01	1.02	1.02	1.01	1.02	1.01	1.01	1.01	1.01	1.00
9	1.02	1.01	1.01	1.02	1.01	1.01	1.02	1.01	1.01	1.01	1.01	1.00
10	1.02	1.01	1.01	1.02	1.01	1.01	1.02	1.01	1.01	1.01	1.01	1.00

Table 3-7   Factors for effective floor cavity reflectances other than 20 percent

Substituting all our known values in Equation 3-4, we now calculate as follows:

$$\text{No. of luminaires} = \frac{150 \times 1350}{0.58 \times 4800 \times 2 \times 0.79 \times 0.90}$$

$$= 48$$

Spacing limitations between semi-direct luminaires are related to the mounting height above the work plane. The recommended spacing-to-mounting-height-above-the-work-plane, in this case, is 1.2 (see Fig. 3-7). So our spacing between rows should not exceed 1.2 × 6 = 7.2 ft. If we place the outside rows 2-1/2 feet from the walls, we find that we need 6 rows to keep from exceeding the maximum spacing between rows. (We can, of course, use somewhat closer spacing between outer rows than between central rows.) From Fig. 3-9, we see that the end-to-end spacing cannot exceed the room cavity height minus 4 feet. This means the end-to-end spacing must not exceed two feet (6 − 4 = 2). With 4-foot fluorescents, eight to a row, spaced 1-1/2 feet apart, and their ends 1-1/2 feet from the walls, means that we use a total of 48 luminaires. For this particular application, the end lighting is likely to be entirely satisfactory since the work plane will probably set in about 2 feet from each end.

3.5   *Current and Power Requirements*

For incandescent lighting systems, the power factor is unity (1) and the total power is simply the sum of the wattage ratings of the individual lamps. The current is simply $I = P/E$, where P is the total power and E is the applied EMF.

The NEC does not require that the power rating of fluorescent units be given. It does require, however, that all such units be plainly marked with their operating voltage, current—including ballasts and transformers—and frequency.

Without correction, the power factor of fluorescent units most often ranges from 0.5 to 0.6. When power-factor correction is provided, by means of a capacitor built into the ballast, the unit is identified in technical data as HPF (high power factor), and the PF is about 0.9.

To illustrate the wisdom of using HPF units, assume the use of a fixture that has two 40 W lamps operating at 115 V, 60 Hz.

The total *lamp* power is 80 W. To allow for losses by auxiliary equipment, add 20 percent. In this case, $0.20 \times 80 = 16$ W, and the total *fixture* power is $80 + 16 = 96$ W.

With a HPF unit,

$$I = \frac{P}{E(\text{PF})} = \frac{96}{115(0.9)} = 0.928 \text{ A}$$

If a non-corrected unit is used,

$$I = \frac{P}{E(\text{PF})} = \frac{96}{115(0.5)} = 1.67 \text{ A}$$

If the lamps were to be installed in a circuit carrying a maximum of 15 A, we could use 16 of the HPF units, but only 8 of the non-corrected units.

Although incandescent lighting fixtures may be less expensive to install, they produce less lumens per watt than fluorescent units. Thus, for a given level of light, operating costs are higher with the incandescent units. In air-conditioned spaces, the use of incandescent fixtures also increases the air-conditioning load.

EXERCISES

3-1/   Name five variables that affect lighting.

3-2/   Define the term *footcandle*.

3-3/   Define the term *lumen*.

3-4/   What relationship exists between footcandles and lumens?

3-5/   What is a *light meter*, and how does it work?

3-6/   Give an example of how light varies with distance.

3-7/   What advantages are associated with incandescent lamps?

3-8/   What blackens the inner surface of an incandescent lamp as the lamp ages?

3-9/   How is an incandescent lamp affected when it is operated at higher-than-design voltages?

3-10/   How is an incandescent lamp affected when it is operated at lower-than-design voltages?

3-11/   What are the advantages of fluorescent lamps compared to those of the incandescent variety?

3-12/   What are the disadvantages of fluorescent lamps compared to the incandescent variety?

3-13/   Explain how a fluorescent tube using *preheat* starting operates.

3-14/   Describe how a *glow-type* starter works.

3-15/   How does a *bimetallic strip* operate?

3-16/   What two functions are performed by a *ballast*?

3-17/   Why is a *capacitor* connected across the starter contact points?

3-18/   Explain how a fluorescent lamp can be operated from a dc source of EMF.

3-19/   Why is a capacitor often placed in series with the filament of one lamp in a multi-lamp fluorescent fixture?

3-20/   Name *three* starting methods for fluorescent lamps and describe each briefly.

3-21/   How are fluorescent lamps rated? How does performance vary with deviations from rated voltages?

3-22/   What is actually meant by the term fixture? What other name is used in some literature in place of fixture?

3-23/   What is the coefficient of utilization?

3-24/   What determines lamp lumen depreciation?

3-25/   Explain the significance of the LLD.

3-26/   What is the power factor in an incandescent lighting system? Why?

3-27/   What is the power factor in fluorescent lighting systems. Why?

3-28/   Explain how the power factor can be improved in a fluorescent lighting system.

3-29/   What practical advantage is associated with the use of HPF fluorescent fixtures?

3-30/   Over a span of time, which type of lighting system, incandescent or fluorescent, would prove the more economical? Why?

Part
2

4

Electrical
Conductors

4.1   *The National Electrical Code*

The *National Electrical Code* (NEC) first adopted in 1897, is sponsored by the
National Fire Protection Association (NFPA) under the auspices of the
American National Standards Institute (ANSI).

To assure the standardization and proper installation of electrical
equipment, the Code establishes *minimum* standards for electrical installations
within buildings and other premises.

The Code is divided into *Articles*, each of which contains as many
Sections (subdivisions) as is necessary for clarification and comprehensiveness.
Each Article or group of Articles is written by a panel of experts. Updates and
changes are made from time to time so that a revised Code is published about
every three years.

The adoption and enforcement of the NEC is usually left to the
discretion of *local* authorities rather than to state or national authorities. Local
wiring inspectors may therefore differ in their interpretation of some provisions
of the Code since NEC standards are *minimum* requirements. In the absence of
any local authority, the NEC prevails, however—usually under the authority of
the electric utility company.

The NEC is *not* an instruction manual, but it does set an official minimum standard for the determination of many electrical design problems.

So that you will be able to use the Code intelligently, take a few moments now to scan its Table of Contents and to read Article 90 the Introduction, in its entirety. Read also Article 100, Definitions, and Article 110, Requirements for Electrical Installations. By so doing, you will acquire a "feel" for the language of the Code.

Notice that a two-number method of identification is used in referring to a particular portion of the Code. The first number identifies the Article and the second number identifies a Section of that particular Article. For example, Reference 90-1, indicates Article 90, Section 1.

A similar identification method is used to refer to sections of this text, with the first number designating the Chapter and the second number designating the Section. Since the lowest-numbered Article in the Code is 90, while the highest numbered chapter in this book is 13, however, no confusion should result.

Product Listing and Labeling

Before starting the production of electrical products, most U.S. manufacturers submit samples of their product to a nationally recognized testing laboratory, inspection agency, or other organization concerned with product evaluation. Those products which meet nationally recognized standards are then *listed* by these organizations. Listing simply means that the items have been found suitable for use in a specified manner.

Some listing organizations also require *labeling*; that is, the equipment must carry a label, symbol, or other identifying mark which indicates compliance with nationally recognized standards or tests to determine suitable usage in a specified manner.

It is always wise to check with a local inspector to see if you should buy listed and/or labeled electrical equipment.

Local codes, permits, and licenses

Frequently, local ordnances limit or qualify the NEC standards. You must, therefore, also be familiar with the ordnances which apply in your region.

Always check with your local governmental agency to determine whether or not a permit is required before a wiring job is started. If one is required, the permit must be shown to the local inspector. Power companies may refuse to supply power in the absence of an inspection certificate.

If electrical wiring is your *business*, state laws usually dictate that you obtain a license. On the other hand, if you are doing wiring in your own home and do not engage in this activity as a business, no license is generally required. To avoid possible legal difficulties, however, verify the *no license* requirement before you start work.

4.2   *The Circular Mil and Wire Sizes*

As we saw in Sec. 1.9 of this text, when current passes through the resistance of any practical electric circuit, the *rate* at which electrical energy is converted to heat energy is called *power*, designated *P*. Mathematically,

$$P = I^2 R \tag{4-1}$$

Equation 4-1 tells us that the amount of heat generated by a current-carrying conductor of fixed resistance is *directly proportional* to the *square* of the current passing through the conductor. Thus, if the current is doubled, the amount of heat the conductor can dissipate to the ambient (surrounding environment) is quadrupled. If the current is tripled, the amount of heat the conductor must dissipate is multiplied by a factor of 9. Obviously, with a conductor of given size, if the current is steadily increased a point is eventually reached where the conductor cannot dissipate the amount of heat generated and a definite fire hazard exists.

As will be shown in Sec. 4.7, the NEC specifies the maximum amperage a given conductor can handle safely under varying conditions. At this point, however, let us turn our attention to the problem of how wire sizes are specified.

Conductors used for residential wiring are usually round, and the unit used for measuring their cross-sectional area is the circular mil, abbreviated cmil. A mil is 0.001 inch and a conductor having a cross-sectional area of 1 cmil is one whose diameter is 1 mil.

The cross-sectional area of any round conductor is, therefore, equal to the square of the diameter in mils; that is,

$$A = D^2 \tag{4-2}$$

where A is the area in cmil and D is the diameter in mils.

If, for example, a round conductor has a diameter of 3 mils (0.003 in), its cross-sectional area is 3^2 or 9 cmil. To take another example, if a round conductor has a diameter of 9 mils (0.009 in), its cross-sectional area is 9^2 or 81 cmil.

The American Wire Gauge

To assure the manufacture of conductors in sizes suitable for all applications, the American Wire Gauge was developed to assign a number to a particular size wire. These AWG numbers start at 40, the *smallest*, having a diameter of 3.145 mils which is about the diameter of a human hair. The gauge numbers then *descend* in order to 0000, the *largest*, with a diameter of 460 mils. The size of any conductor larger than 0000 is expressed directly in its circular-mil area, beginning with 250,000 cmil. The largest recognized standard size is 2,000,000 cmil.

In some literature the reader will find circular mil abbreviated CM and the prefix M used to indicate thousand. Thus, 250,000 cmil, for example, may also be written as 250 MCM. It is also common practice to designate size 00 as 2/0, size 000 as 3/0, and size 0000 as 4/0.

Table 4-1 lists the properties of standard-size conductors from #18 to 1-million cmil. Except as provided by NEC Sec. 310-3, all conductors #8 and larger are stranded.

AWG size	Area (cmil)	No. of wires	dc resistance at 25°C, 77°F (Ω/1000 ft)	
			Copper	Aluminum
18	1620	Solid	6.51	10.7
16	2580	Solid	4.10	6.72
14	4110	Solid	2.57	4.22
12	6530	Solid	1.62	2.66
10	10,380	Solid	1.018	1.67
8	16,510	Stranded	0.6404	1.05
6	26,240	Stranded	0.410	0.674
4	41,740	Stranded	0.259	0.424
3	52,620	Stranded	0.205	0.336
2	66,360	Stranded	0.162	0.266
1	83,690	Stranded	0.129	0.211
0	105,600	Stranded	0.102	0.168
00	133,100	Stranded	0.0811	0.133
000	167,800	Stranded	0.0642	0.105
0000	211,600	Stranded	0.0509	0.0836
	250,000	Stranded	0.0431	0.0708
	300,000	Stranded	0.0360	0.0590
	350,000	Stranded	0.0308	0.0505
	400,000	Stranded	0.0270	0.0442
	500,000	Stranded	0.0216	0.0354
	600,000	Stranded	0.0180	0.0295
	700,000	Stranded	0.0154	0.0253
	750,000	Stranded	0.0144	0.0236
	800,000	Stranded	0.0135	0.0221
	900,000	Stranded	0.0120	0.0197
	1,000,000	Stranded	0.0108	0.0177

Table 4-1   Properties of conductors

The last two columns of Table 4-1 give the dc resistance in ohms per 1,000 feet (Ω/1000 ft) of copper and aluminum conductors at room temperature—25°C or 77°F. The practical application for information of this type is shown in Sec. 4.9 of this chapter.

A *wire gauge,* which is simply a hand-held, circular metal disk with slots around its edge, is handy for determining the exact size of a wire. Each slot is marked with its corresponding AWG number. The slot into which the wire fits, *not* the opening behind the slot, indicates the wire size.

Skin effect

When ac flows in a conductor it has a tendency to flow toward the outside of the conductor rather than through the cross section uniformly. This so called *skin effect* increases as the frequency of the ac is increased. Its effect is slight in small conductors but it must be carefully considered in large conductors. To compensate for skin effect at a frequency of 60 Hz ac, multiply the dc resistance shown in Table 1-4 by the appropriate factor listed in Table 4-2.

AWG Size or cmil	For Nonmetallic-Sheathed Cables or in Air		For Metallic-Sheathed Cables or in Metallic Raceways	
	Copper	Aluminum	Copper	Aluminum
Up to 3	1	1	1	1
2	1	1	1.01	1
1	1	1	1.01	1
0	1.001	1	1.02	1
00	1.001	1.001	1.03	1
000	1.002	1.001	1.04	1.01
0000	1.004	1.002	1.05	1.01
250,000	1.005	1.002	1.06	1.02
300,000	1.006	1.003	1.07	1.02
350,000	1.009	1.004	1.08	1.03
400,000	1.011	1.005	1.10	1.04
500,000	1.018	1.007	1.13	1.06
600,000	1.025	1.010	1.16	1.08
700,000	1.034	1.013	1.19	1.11
750,000	1.039	1.017	1.21	1.12
800,000	1.044	1.017	1.22	1.14
1,000,000	1.067	1.026	1.30	1.19

Table 4-2   Multiplying factors for converting dc resistance

Table 4-2 shows that for sizes smaller than 2/0, skin effect requires little consideration, but for large sizes it is important, especially when conductors are enclosed in metal *raceways*, that is any channel for holding wires, cables or bus-bars which is designed expressly for and used solely for this purpose. Notice also that the correction factor for copper is appreciably *higher* than for aluminum.

Wire-gauge approximations

In setting up the AWG wire table, the cross-sectional dimensions are chosen so that each *decrease* of one gauge number represents a 25 percent *increase* in cross-sectional area. Thus, a decrease of three gauge numbers represents an increase in cross-sectional area of 1.25 × 1.25 × 1.25 or approximately a 2:1 increase. Similarly, a change of ten wire-gauge numbers represents a change of 10:1 in cross-sectional area. Moreover, since doubling the cross-sectional area cuts the resistance in half, a decrease of three wire-gauge numbers cuts the resistance of a conductor of a given length in half. If we memorize the dimensions for any one wire-gauge number, therefore, we can learn the approximate dimensions for any other wire-gauge number. For this purpose, AWG #10 wire is quite convenient. For practical purposes, the diameter is taken as 100 mil, the area as 10,000 cmil, and the resistance for copper as 1 Ω/1000 ft at room temperature.

To illustrate the usefulness of the above approximations, let us determine the resistance of 200 ft of AWG #14 copper wire at room temperature.

Since the cross-sectional area of AWG #14 wire is $1/(1.25)^4$ that of AWG #10 wire,

$$R = 2.5 \times \frac{200}{1000} \times 1\ \Omega = 0.5\ \Omega$$

4.3   *The NEC Color Scheme*

Any special exceptions will be discussed as the need arises, otherwise the general color scheme established by the NEC is explained below.

Wires

1.   Only *white* may be used for the grounded neutral wire.

2.   The grounding wire from a normally non-current carrying component, such as the frame of a motor, may be *green, green and yellow stripe,* or *bare* (uninsulated). The green and yellow stripe is usually found on those appliances intended for export to European markets as well as those for domestic sale.

3.   The color of the *hot* wire (or wires) used in conjunction with the white neutral wire depends on the number of wires in the circuit and is as follows:

2-wire circuit — black

3-wire circuit — black, red

4-wire circuit — black, red, blue

5-wire circuit — black, red, blue, yellow

Terminals

1.   On *grounding* receptacles, that is, two parallel slots plus a third round or U-shaped opening for a corresponding prong on the plug, the round or U-shaped opening leads to a special *green* terminal. When installing the receptacle, a wire (green, green and yellow stripe, or bare and uninsulated) must be connected from the green terminal to ground.

2.   On switch lugs, receptacles, and the like, the "hot" terminal is *natural copper* or *brass* and the "ground" terminal is a *whitish* color, such as *nickel, tin,* or *zinc plate.*

4.4   *Insulation*

The term *wire* generally refers to any *round* conductor up to AWG #4/0. A conductor larger than size 4/0 is a *cable.* Both wires and cables are distinguished from flat bars, called *bus bars*, which are often used in place of cables of one million cmil and larger. Due to its increased rectangular surface area, a bus bar provides better cooling than round cable and is usually less cumbersome to install.

The *lightest* size wire permitted by the NEC for ordinary residential wiring is size AWG #14. Sizes greater than 4/0 are seldom encountered in such applications.

Locations

Wires of a given type are often restricted to use in locations referred to as *damp, dry,* and *wet.* Article 100 of the Code defines these three types of locations.

Dry locations — A location classified as dry and not normally subject to dampness or wetness may be temporarily subject to dampness or wetness, however, as is the case when a building is under construction.

Damp location — A partially protected location under canopies, marquees, roofed open porches, and the like, as well as interiors subject to moderate degrees of moisture—some basements and bars, and some cold-storage warehouses.

Wet location — Installations underground or in concrete, slabs or masonry in direct contact with the earth, and any locations subject to saturation with water or other liquids (vehicle-washing areas, for example) and unprotected locations exposed to weather.

Insulation

For many years, *natural rubber* was the conductor insulating material most widely used. Nevertheless, such insulation dried out, cracked, and became brittle from age and heat. So now rubber has been largely replaced with *thermoplastic*. Other improved materials are also being developed constantly. In 1968, for example, the Code first recognized a *cross-linked synthetic polymer* which has insulating characteristics far superior to either rubber or thermoplastic. It may very well replace the other two almost entirely in the years ahead.

The amount and type of insulation to be utilized is determined by the potential difference (voltage) between conductors and by the location (damp, dry, or wet) in which the conductor is to be used.

To identify the numerous types of insulation, the NEC has established a letter code system which describes their characteristics. Some of the most common types still being manufactured are shown in Table 4-3. The letters, R, T, and X refer to the material—rubber, thermoplastic, or cross-linked synthetic polymer. H and HH will endure progressively higher temperatures. W indicates that a given insulation is moisture resistant, while N indicates a final extruded jacket of nylon.

The insulations listed in Table 4-3 are suitable for ordinary residential and farm wiring at 600 volts or less. A complete listing of conductor insulations, maximum operating temperatures, application provisions, thickness of insulation, and type of outer covering is given in NEC Table 310-13.

4.5   *Cables*

In the preceding section, any conductor larger than AWG #0000 was classified as a cable. The word *cable* is also generally taken to mean two or more insulated conductors enclosed by the manufacturer in a common external jacket which may be either metallic or nonmetallic. In referring to such cables it is common to use a designation such as 12-4, for example; the 12 indicating wire size and the 4 the number of wires in the cable.

---

*   Reprinted with permission from the 1975 National Electrical Code, Copyright 1975, National Fire Protection Association, Boston, Ma.

Trade Name	Type Letter	Maximum Operating Temperature °F	Application
Heat resistant	RHH	194	Dry locations
Mositure and heat-resistant rubber	RHW	167	Dry and wet locations
Thermoplastic	T	140	Dry locations
Moisture-resistant thermoplastic	TW	140	Dry and wet locations
Moisture- and heat-resistant thermo-plastic	THW	167	Dry and wet locations
Moisture- and heat-resistant thermo-plastic	THWN	167	Dry and wet locations
Heat-resistant thermoplastic	THHN	194	Dry locations
Moisture- and heat-resistant cross-linked synthetic polymer	XHHW	194	Dry locations
		167	Wet locations

Table 4-3   The NEC letter system for insulation identification

Metal-clad cable

Two series of metal-clad cable, MC and AC, are permitted by the Code (Article 334). Cables of the MC series are *power* cables with conductors of size AWG #14 and larger for copper and AWG #12 for aluminum. The conductor enclosure is either interlocking metal tape, or an impervious, close fitting, corrugated tube. Series MC cables are not used in ordinary residential wiring.

Cables of the AC series, commonly known as BX cables, are frequently used in residential wiring. The covering for these cables is a galvanized steel spiral wrapping. Where the cables terminate, the conductors must be protected from the sharp edges of the armor by proper bushings and fittings.

For rubber-covered conductors, the cable is designated type AC. For thermoplastic-covered conductors, the cable is type ACT. The use of AC and ACT cables is restricted to dry locations. Type ACL cable, containing lead-covered conductors, is used for damp or wet locations.

Cables of the AC series are available with either two or three conductors. In keeping with the NEC color scheme, the conductors of the two-wire variety are black and white, while those of the three-wire variety are

---

* Reprinted with permission from the 1975 National Electrical Code, Copyright 1975, National Fire Protection Association, Boston, Ma.

black, white, and red. In either variety, there is also a bare, uninsulated copper strip or wire for the ground circuit.

Nonmetallic sheathed cable

A nonmetallic sheathed cable is an assembly of two or more insulated conductors having an outer sheath of moisture-resistant, flame-retardant, nonmetallic material (Sec. 336-1).

Two types of nonmetallic sheathed cable are permitted; NM and NMC. The use of type NM, often called *Romex,* is restricted to normally dry locations. For damp or wet locations, type NMC must be used.

Both types are available with either two or three conductors and most also contain a bare, uninsulated wire for grounding.

Conductor sizes range from No. 14 to 2 AWG for copper, and from No. 12 to 2 AWG for aluminum and/or copper-clad aluminum.

Service-entrance cable

Types SE and USE only are permitted. As the name implies, cables of this type are used primarily for service entrances, but they may also be used as specified for certain other purposes (Sec. 338-3). The prefix U indicates underground. These cables will be thoroughly described in Chapter 6 of this text.

Miscellaneous cable types

Type MI or mineral-insulated metal-sheathed cable (Article 330), Type ALS or aluminum-sheathed cable (Article 331), Type SNM or shielded nonmetallic sheathed cable (Article 337), and Type UF underground feeder and branch-circuit cable (Article 339) are all cables not ordinarily installed in normal residential wiring, but unusual circumstances may dictate their use.

4.6   *Wire Miscellany*

Fixture wire

This is a special type of wire used only for the internal wiring of lighting fixtures; it is never used in a circuit leading up to a fixture. The minimum size for fixture wire is #18 AWG.

Numerous types of fixture wire are available. The type you select for a particular application is determined primarily by the circuit amperage and ambient temperature.

To avoid a rather lengthy discussion of limited value in this text, the reader is referred to Article 402 of the Code.

Low-voltage wire

When a low potential, usually 30 volts or less, is applied to a circuit, *bell* wire (also called *annunciator* wire) is used. The conductors are No. 18 for residential wiring and the insulation is paraffin-impregnated cotton. Common applications include doorbells and thermostats.

4.7 *Ampacity*

The maximum current in amperes that a conductor can carry safely and continuously is called its *ampacity*. In addition to its cross-sectional area, there are several factors which affect the ampacity of a given conductor. If conductors are in free air where they can readily dissipate heat, their ampacity is much greater than if enclosed in a raceway—the outer sheath of cables is considered a raceway in this context. Moreover, when more than three conductors are in the same raceway, their ampacity is systematically reduced so that the heat of the ambient, plus that produced by the passage of current through the conductors, does not exceed the heat rating of the conductor insulation.

NEC Code Tables 310-16 through 310-19, and their associated notes,

	Copper		Aluminum	
Size AWG	140°F (60°C) R, T, TW	167°F (75°C) RH, RHW, TH, THW	140°F (60°C) R, T, TW	167°F (75°C) RH, RHW TH, THW
14	20	20		
12	25	25	20	20
10	40	40	30	30
8	55	65	45	55
6	80	95	60	75
4	105	125	80	100
3	120	145	95	115
2	140	170	110	135
1	165	195	130	155
0	195	230	150	180
00	225	265	175	210
000	260	310	200	240
0000	300	360	230	280

Table 4-4   Allowable ampacities in amperes of single, insulated conductors in free air

provide detailed information on the maximum, continuous ampacities of copper and aluminum conductors.

Tables 4-4 through 4-6 shown below are excerpts from those Code Tables. The following examples will show you some practical application of these tables.

Size AWG	Copper		Aluminum	
	140°F (60°C) R, T, TW	167°F (75°C) RH, RHW TH, THW	140°F (60°C) R, T, TW	167°F (75°C) RH, RHW TH THW
14	15	15		
12	20	20	15	15
10	30	30	25	25
8	40	45	30	40
6	55	65	40	50
4	70	85	55	65
3	80	100	65	75
2	95	115	75	90
1	110	130	85	100
0	125	150	100	120
00	145	175	115	135
000	165	200	130	155
0000	195	230	155	180

Table 4-5   Allowable ampacities of insulated conductors, maximum of three in raceway or cable—based on ambient temperature of 86°F (30°C) and expressed in amperes

Ambient Temp.		Correction Factor For Conductors			
°F	°C	140°F 60°C	167°F 75°C	185°F 85°C	194°F 90°C
104	40	.82	.88	.90	.90
113	45	.71	.82	.85	.85
122	50	.58	.75	.80	.80
131	55	.41	.67	.74	.74
140	60	....	.58	.67	.67
158	70	....	.35	.52	.52
167	75	....	....	.43	.43
176	80	....	....	.30	.30

Table 4-6   Correction factors—ambient temperature higher than 86°F (30°C) for conductors approved for indicated temperatures

Let us determine the ampacity of size AWG #8 copper conductors with THW insulation, three in a cable. The installation is in a location where the ambient temperature is expected to reach a maximum of 104°F (40°C).

In Table 4-5 we see that the temperature rating of the conductor is 167°F (75°C) and the ampacity, based on an ambient temperature of 86°F (30°C) is 45 A. In Table 4-6, the derating factor at the elevated temperature is shown as 0.88. Thus, the ampacity is

$$45 \times 0.88 = 39.6 \text{ A}$$

As another example, suppose that for a load rated at 30 A you want to know what size of copper conductors (Type THW insulation) to use in a three-wire cable. Again, let us presume an ambient temperature reaching a maximum of 104°F (40°C).

From Table 4-6 we learn that the derating factor for 104°F is 0.88 when using copper conductors with Type THW insulation. Thus, the ampacity must be increased to

$$30/0.88 = 33.1 \text{ A}$$

By checking Table 4-5, we can then see that an ampacity of 33.1 A requires three size AWG #8 conductors in a cable.

Where the number of conductors in a raceway or cable exceeds three, additional derating (reduction) of ampacity is necessary. Setting aside the exceptions which the Code identifies in the notes for Tables 310-16 through 310-19, the required reductions are listed in Table 4-7.

Number of conductors	Percent of values in NEC Tables 310-16 and 310-18
4 to 6	80
7 to 24	70
25 to 42	60
43 and above	50

Table 4-7   Reduction factor applied to ampacities when the number of conductors in a raceway or cable exceeds three

Any exceptions to the above procedures for determining ampacity are discussed as the need arises.

4.8   *Temperature Conversions*

At times, it may be very convenient to know how to convert °F to °C or vice versa.

To express °F in °C, use the relationship

$$°C = \frac{5}{9}(\text{Temp. in } °F - 32) \tag{4-3}$$

For example, we convert 140°F to the equivalent °C as follows:

$$°C = \frac{5}{9}(140 - 32)$$

$$= \frac{5}{9} \times 108$$

$$= 60$$

To convert °C to °F use the relationship

$$°F = \frac{9}{5}(°C) + 32 \tag{4-4}$$

To express 60°C in °F, for example, we have

$$°F = \frac{9}{5}(60) + 32$$

$$= 108 + 32$$

$$= 140$$

4.9   *Voltage Drops*

Two basic problems influence the selection of the *correct conductors* for a given application; namely, *ampacity,* which we considered in Sec. 4.7, and *voltage drop.*

In Sec. 1.9, we learned that power is *directly proportional* to the *square* of the voltage drop appearing across the resistance of any practical conductor ($P = V^2/R$). Since power represents the rate at which electrical energy is converted to heat energy, and since the heat loss performs no useful function, it represents *waste.* As you well know from your monthly utility bill, electrical energy is too expensive to waste.

An equally important reason for minimizing such waste is that the quality performance of electrical devices is usually dependent on the voltage they receive. Suppose, for example, a lighting system, as a result of the voltage drop in the conductor connecting it to the source, operates 95 percent of its rated voltage. Since

$$P = \frac{V^2}{R} = (0.95)^2 = 90 \text{ percent}$$

the lights will produce only 90 percent of their rated output.

From the above, it is apparent that we must select conductors of cross-sectional area sufficient to keep the voltage drop within satisfactory limits. Of course, the wire selected must also have sufficient ampacity. When voltage drop and ampacity are both considered, select the *largest* conductor.

Section 210-19(a) of the Code reads as follows: "Conductors for branch circuits as defined in Article 100, sized to prevent a voltage drop exceeding 3 percent at the farthest outlet of power, heating, and lighting loads, or combinations of such loads and where the maximum total voltage drop on those feeders and branch circuits to the farthest outlet does not exceed 5 percent, will provide reasonable efficiency of operation."

In ordinary residential wiring, *branch circuits* start at the fuse (or circuit-breaker) panel and there are no *feeders*. Thus the voltage drop in the system should not exceed 3 percent of the supply voltage. On farms, feeders run from a pole-mounted meter to the point wehre they enter a building. (Complete definitions for the terms branch circuit and feeder are given later in Sec. 6.1 of this text.) Allowing a 2 percent drop on the feeders plus a 3 percent drop on branch circuits, the overall drop is 5 percent. Notice, however, that the recommended voltage drops are *maximums*. In practice, a 2 percent drop on both branch circuits and feeders is a wise—widely accepted—choice when the cost of the heavier conductors is weighed against the cost of power loss in smaller-size conductors over an extended period of time.

To explain selection of satisfactory conductors, let us presume the use of copper conductors with Type T insulation and in a metallic raceway to supply an 80 A load: the load to be located 150 feet from a single-phase, 115 volt 60 Hz ac source; the voltage drop not to exceed 2 percent of the source voltage; and the ambient temperature never to rise above 86°F (30°C).

Given these specifications, the voltage drop must not exceed 0.02 × 115 = 2.3 V. Since the load is 150 feet from the source, the total length of wire required is 2 × 150 = 300 feet. The resistance of the conductor, by Ohm's law, is therefore

$$R = \frac{V}{I} = \frac{2.3}{80} = 0.02875 \ \Omega \text{ for 300 ft}$$

Table 4-1 in Sec. 4-2 showed the dc resistance in ohms per 1,000 feet for copper conductors. To determine which size wire will limit the voltage drop to 2.3 V over a distance of 300 feet we use the relationship

$$\Omega/1000 \text{ ft} = \Omega \text{ for actual distance} \times \frac{1000}{300}$$

$$= 0.02875 \times 3.33$$

$$= 0.09574 \ \Omega$$

From Table 4-1, we find that the wire having this value of resistance per 1000 feet is size AWG #2/0.

To compensate for skin effect, we refer to Table 4-2 and use the multiplying factor 1.02. Then,

$$\Omega/1000 \text{ ft} = 0.09574 \times 1.02 = 0.0976 \ \Omega$$

Even after compensation for skin effect, size AWG #2/0 is still the correct choice.

Referring to Table 4-5 we now find that AWG #2/0 wire has an ampacity of 145 A, which is entirely satisfactory for this application.

Notice in Table 4-5 that, *without considering the voltage drop,* the selection of size AWG #3 is indicated; and in Table 4-1, the dc resistance per 1000 feet of #3 is 0.205 Ω. Each 100 feet would, therefore, have a resistance of 0.0205 Ω and the resistance for 300 feet would be 0.0615 Ω. Using these values the voltage drop across the conductors would be

$$V = IR = 80 \times 0.0615 = 4.92 \text{ V}$$

which is more than twice the allowable voltage drop. This simply reinforces the earlier statement that, considering both the ampacity tables and the voltage drop, the largest conductor is the one to use.

Now let us take a look at another, more direct method for determining the wire size to use for a particular application. The equation to use is

$$A = \frac{ID\rho}{V} \qquad\qquad (4\text{-}6)$$

where A is the cross-sectional area in circular mils, V is the allowable voltage drop across the conductors, *I* is the load current in amperes, D is the distance from the supply to the load in feet, and ρ (the Greek letter rho) is the so-called *resistivity* of the conductors.

By definition, resistivity is simply the resistance in ohms of a sample 1 foot long and 1 mil in diameter. The unit of resistivity is ohms-cmils per foot (Ω-cmil/ft).

At first glance, the introduction of resistivity seems to throw a very complicated factor into otherwise simple calculations. Fortunately, this is not so. For electrical work all you have to remember is that the resistivity of copper is 10.4 Ω-cmil/ft and for aluminum, 17 Ω-cmil/ft.

Considering only the right-hand member of Equation 4-6, we have the units

$$\frac{A \cdot ft \cdot \frac{\Omega\text{-cmil}}{ft}}{V}$$

In the numerator, the unit feet cancels out leaving

$$A \cdot \frac{\Omega\text{-cmil}}{V}$$

Since $V = IR$ we can rewrite the above as

$$\frac{A \cdot \Omega\text{-cmil}}{A\Omega}$$

When A and Ω are canceled out we are left with cmil, the unit required for area in Equation 4-6.

Before we can work with Equation 4-6, one other factor, distance D, requires special comment.

For single-phase, two- or three-wire systems, distance D must be multiplied by *two*. The neutral conductor of a single-phase, three-wire system is *not* included since it carries no current with a balanced load and, therefore, creates no voltage drop. A three-wire system is designed so that it is balanced when the total load is applied. If part of the load is not in use, there will be some current through and voltage drop across, the neutral conductor; but for practical computations we set aside this drop and assume a balanced load.

In three-phase systems, distance D is multiplied by the factor 1.732 which is the square root of 3.

To illustrate the usefulness of Equation 4-6, let us use the same data as in the preceding example and see how the two answers compare. For convenience, the data are repeated here.

We want to use copper conductors with Type-T insulation in a metallic raceway to supply an 80 A load. The load is located 150 feet from a single-phase, 115 volt 60 Hz ac source, the voltage drop does not exceed 2 percent of the source voltage, and the maximum ambient temperature is 86°F (30°C).

The voltage drop is 115 × 0.02 = 2.3 V, the resistivity of copper is 10.4 Ω-cmil/ft, I = 80 A, and D = 2 × 150 = 300 ft. Substituting these values in Equation 4-6 we obtain

$$A = \frac{80 \times 300 \times 10.4}{2.3}$$

$$= 108,522 \text{ cmils}$$

From Table 4-1, size AWG #00 has an area of 133,100 cmil and is the wire selected. This is in agreement with the more roundabout selection technique demonstrated earlier.

In practice, economic considerations may dictate the acceptance of a somewhat higher voltage drop. That would be the case, for example, if the load in the above example was used intermittently and for short periods of time. The cost of any power wasted over a good many years under such circumstances would probably be much less than the cost of the heavier wire needed for a 2 percent voltage drop. In other words, common sense plays a very important role in the specifications for any actual installation.

Where it is permissible, the use of 230 volts instead of 115 volts may also result in an appreciable materials savings and, at the same time, keep the voltage drop within acceptable limits.

To illustrate the validity of the above statement, recall that $P = EI$. For a given power, if the voltage is doubled, the current is halved. With the current halved, the voltage drop across the fixed resistance of a given length of a particular size wire is also halved since $V = IR$. If the voltage drop was 2 percent of the applied EMF with a 115 volt source, it is reduced to 1 percent when we use a 230 volt source. This means that we can use a smaller-size wire and still limit the voltage drop to 2 percent of the applied EMF.

In the previous example, we used a 115 volt source and found the required wire size to be 2/0. Since the voltage drop across the conductors was limited to 2.3 volts, the voltage drop across the source was $115 - 2.3 = 112.7$ volts. The power dissipated by the load is

$$P = EI = 112.7 \times 80 = 9016 \text{ W}$$

Using a 230 V source, we can now determine the size of the wire required to deliver the same 9016 W to the load. We again use Equation 4-6, but now, I is reduced from 80 to 40 A and the allowable voltage frop is $230 \times 0.02 = 4.6$ V. Therefore,

$$A = \frac{40 \times 300 \times 10.4}{4.6}$$

$$= 27{,}130 \text{ cmils}$$

There is a big difference between the cost of size AWG 2/0 and size AWG 4 which has an area of 41,740 cmils. See Table 4-1.

4.10   *Aluminum Wire*

It was noted earlier that the resistivity of aluminum is 17 Ω-cmil/ft while that of copper is 10.4 Ω-cmil/ft. This means that the conductivity of aluminum is considerably *lower* than that of copper, so a *larger* size aluminum conductor is needed to carry a given current. For ordinary wiring, aluminum conductors are *usually* two AWG sizes larger than would be necessary for copper. This specification is *not always* accurate, however. Always check the ampacities of aluminum conductors by referring to Tables 310-18 and 310-19 of the Code.

Because copper is, at present, much more common than aluminum in residential wiring, this text is concerned mainly with copper.

Major drawbacks to the use of aluminum conductors in residential wiring include the relatively great difficulty of soldering it, its tendency to loosen at screw terminals, and the present limitations on the use of solderless connectors.

4.11   *Raceways and Wireways*

As we noted earlier, a raceway is a channel for holding wires, cables, or bus bars. For cables, the raceway is an integral part of the complete assembly. Here, however, we are concerned with raceways which are completely separate from the conductors. Included in this group are *conduit, electrical metallic tubing* (EMT), *surface raceways,* and *wireways.*

Raceways provide mechanical protection for the conductors and are mechanically installed as a complete system with all necessary outlet boxes and fittings. Conductors are then pulled through the raceway.

Rigid metal conduit (Article 346)

In outward appearance, rigid metal electrical conduits look like the pipes used for water, gas, or steam. However, they are lined with a smooth enamel which makes it easy to pull wires through them. Manufactured in 10 foot lengths, such rigid metallic conduit is connected to junction, outlet, and switch boxes, by means of locknuts and bushings. This sort of conduit is generally used where there is a strong chance of mechanical injury or where moisture presents a problem.

Flexible metal conduit (Article 350)

Generally called *greenfield,* flexible metal conduit is, in effect, the outer casing only of Type AC armored cable. Its use is prohibited in wet locations. Generally, the smallest size is 1/2″ electrical trade size. A separate grounding wire is required when conductors are pulled through the conduit. Use of greenfield is rather limited, but it may prove convenient when an occasional change in the location of the load is required. (*The term electrical trade size does not denote the actual physical dimension.* See Table 4, Chapter 9 of the Code.)

Electrical metallic tubing (Article 348)

The EMT-type raceway has a thin wall that does not permit threading. Compression rings or set screws are used to secure connectors and couplings. Tubing sizes range from a minimum of 1/2″ electrical trade size to a maximum of 4″ electrical trade size. The use of EMT is prohibited where it is subject to severe physical damage, in cinder concrete or in fill where it is subject to permanent moisture unless protected on all sides by a layer of noncinder concrete at least 2″ thick or unless the tubing is at least 18″ under the fill. The use of dissimilar metals may cause galvanic action and is not recommended. EMT, like rigid conduit, is manufactured in 10 foot lengths.

Surface raceways (Article 352)

Surface raceways may be either metal or nonmetallic troughs into which the conductors are installed. They are intended primarily for adding extensions to existing circuits rather than for wiring a complete building. The location must be dry and the raceway must be exposed. Elbows, couplings, and similar fittings are so designed that the sections can be mechanically and, if made of metal, electrically coupled while protecting the wiring from abrasions. Holes inside the raceway are drilled so that the heads of screws or bolts are flush with the surface.

Wireway (Article 362)

Wireways are sheet-metal troughs with hinged or removable covers. Conductors are laid in place after the wireway has been installed as a complete unit. Wireways may be installed only for open work; those intended for outdoor use must be made rainproof.

4.12   *Terminal Connections et al.*

Terminal connections (Sec. 110-14)

First the end is prepared by removing the insulation from the wire which attaches to a terminal and by cutting the insulation at a 60-degree angle. This reduces the possibility of nicking and weakening the conductor. Then the insulation is pulled off the conductor and the conductor is cleaned thoroughly.

To attach the conductor to a screw-type terminal, the prepared wire end is bent into a loop, with the loop left wide enough for open insertion under the screw head. Next the loop is inserted under the screw head in such a way that it will close when the screw is tightened. Use long-nosed pliers to close the loop. Then tighten the screw. Correctly done the insulation should nearly butt on the edge of the terminal.

Many modern devices do not have terminal screws. For a good permanent connection it is merely necessary to insert the bare conductor to a depth indicated by a marker on the device. Again, the bare portion of the conductor should extend no more than a fraction of an inch out of the terminal hole.

Solderless terminals with a clamping nut are also commonly used for size AWG #6 and heavier conductors. The stripped and cleaned conductor is simply inserted into the terminal and secured in place by tightening the nut.

Joints

Where joints (splices) are permitted, they must be mechanically and electrically as sound as a continuous conductor; and insulation equivalent to that removed must be used to cover the joint.

When you strip insulation in order to make a joint, taper the remaining insulation to resemble a sharpened pencil.

With solid or small stranded wires, join the two stripped ends at about two-thirds their length. Then wrap the free ends in opposite directions until the joint lies as flat as possible. Butt adjacent turns of the joint and make sure that the free ends terminate at a point close to the original insulation.

With larger stranded wires, bend the individual strands at about their halfway point so that they resemble a cone. Next arrange the cones so that they intermesh. Then take one strand and wrap around the assembly in a clockwise direction. Finally take an opposite strand and wrap it around the assembly in a counterclockwise direction. Make all wraps as tight as possible and with adjacent turns butting. Repeat the procedure with each set of strands until all are wrapped.

Taps

A tap is formed when insulation is removed from a section of continuous wire and a tap wire is joined to the exposed section.

With solid or smaller size stranded wires, place the exposed end of the tap wire *under* the continuous wire. Next, twist the tap wire one full turn in a counterclockwise direction and pass it under itself. Then, wrap the remaining length of tap wire around the continuous wire in a clockwise direction.

If you prefer, you can also place the tap wire *over* the continuous wire initially. Reverse the directions of the wraps from those previously described.

With larger stranded wires, separate the strands of the continuous wire into two equal groups and insert the tap wire into the opening. Then divide the strands of the tap wire into two equal groups and wrap each group around the continuous wire in opposite directions.

Soldering

Once a joint or tap is completed, solder it carefully. Use rosin-core solder—never use acid-core solder.

Before you use the soldering iron, make sure that its tip is clean and well *tinned*; that is covered with a thin coat of solder. In soldering, hold the iron *under* the joint or tap and heat the conductors until the solder applied to the *opposite* side of the conductors flows freely. Do not use an excessive amount of solder; and *do not move* the joint while the solder is hardening. When hard, the solder should have a shiny appearance. If the hardened solder has a dull-gray appearance, resolder the joint or tap.

Improperly soldered connections will exhibit high electrical resistance and will be weak mechanically. Soldering is an art that you can master only with practice. Also, make sure you use an iron of the correct wattage. An iron of low wattage may not provide sufficient heat while one of excessive wattage may damage insulation. If you must use a torch-heated iron, make sure you have proper instruction in its use before attempting any finish work.

New insulation

The final step in making a joint or tap is to reapply equivalent insulation to the exposed conductors. Plastic electrical tape is recommended for this purpose. Start on top of the original insulation at one end and wrap the tape diagonally so that successive wraps overlap slightly. Keep the tape stretched so that successive wraps will fuse to each other. From the opposite end work back to the start with the layers laying at approximately right angles to each other. Repeat until the tape is as thick as the original insulation.

Soldering lugs

First tin the exposed conductor carefully. Then melt a quantity of solder into the lug and while it is still in liquid form, insert the previously tinned conductor. Make absolutely certain the lug attachment is mechanically sound. Limitations on the use of soldering lugs are noted in NEC Code references 230-72 and 250-113.

Solderless connectors

Solderless (pressure) connectors are, in general, intended for applications where there is no strain on a connection. In one type, the stripped conductor ends are simply twisted together. Called a fixture joint it has an insulated shell which is screwed over the twisted wires. Another type has a brass insert and set screw. The wires are clamped in the insert by the set screw and an insulating cap is screwed on over the insert. Another type, handy for making a tap to a continuous wire, has a body with a U-shaped cross section and a nut that slips over the threaded leads of the U.

Special connectors, marked AL for aluminum or CU/AL for copper/aluminum, are required when working with aluminum conductors.

EXERCISES

4-1/   Why is an understanding of the NEC important to anyone undertaking electrical wiring?

4-2/   Does the NEC take preference over local ordinances? Explain.

4-3/   Does the NEC set maximum or minimum standards?

4-4/   What is the function of a nationally recognized testing laboratory?

4-5/   Does a testing laboratory approve electrical items? Explain.

4-6/   What is the area of a round conductor having a diameter of 100 mils?

4-7/   What size conductors are covered for ordinary wiring by the American Wire Gauge? How are these sizes designated?

4-8/   What is a wire gauge, and how is it used?

4-9/   As applied to metallic conductors, what is meant by the expression *skin effect*? How does it affect the resistance of a conductor?

4-10/   Describe the NEC color scheme for (*a*) wires and (*b*) terminals.

4-11/   In terms of size, what is meant by (*a*) wire and (*b*) cable?

4-12/   What is the lightest size wire permitted by the NEC for ordinary residential wiring?

4-13/   What does the NEC mean by (*a*) dry, (*b*) damp, and (*c*) wet locations?

4-14/   Describe a wire insulation designated (*a*) THHN and (*b*) RHW.

4-15/   What is the maximum permissible operating voltage for the insulations described in Sec. 4.4?

4-16/   What does the cable designation 10-6 indicate?

4-17/   What types of metal-clad cable does the Code permit? Which type is quite common in residential wiring and by what name is it usually called?

4-18/   What is meant by the cable designations AC, ACT, and ACL, and in what types of locations can they be used?

4-19/   What color insulations will be found in a three-conductor BX cable?

4-20/   What types of nonmetallic sheathed cable does the Code permit?

4-21/   What location restrictions, if any, are placed on *Romex*?

4-22/   What is meant by the term *ampacity*?

4-23/   What factors affect the ampacity of a conductor?

4-24/   What is the maximum temperature rating of insulation type THHN?

4-25/   What factors must always be considered when selecting conductors for a given application?

4-26/   What are the maximum voltage drops the NEC recommends for (*a*) branch circuits and (*b*) feeders?

4-27/   In general, how does the size of aluminum conductors compare to that of copper conductors for a given application?

4-28/   Define the terms (a) *conduit*, (b) *electrical metallic tubing,* (c) *surface raceway,* and (d) *wireway.*

4-29/   What is *greenfield*?

4-30/   In what lengths are EMT and rigid electrical conduit manufactured?

4-31/   Does the term electrical trade size denote actual physical dimensions? If not, give three examples to illustrate the difference.

4-32/   How is the insulation removed from a conductor to permit its attachment to a screw-type terminal?

4-33/   How is a joint made using large stranded wires?

4-34/   What is a tap and how is one made using solid wires?

4-35/   What precautions must be taken in soldering either a joint or a tap?

4-36/   Using the approximation technique described in Sec. 4.2 of this text, determine the resistance of 500 feet of size AWG 6 copper wire at room temperature.

4-37/   Using either the tables contained in Sec. 4.7 or Code Tables 310-16 through 310-19, as required, determine the ampacity of size AWG 12 conductors with THHN insulation, three in a cable. The maximum ambient temperature is $104°$F.

4-38/   A load takes 48 A. What size copper conductors (Type RHW insulation) must be used in a 3-wire cable when the maximum ambient temperature is $110°$F?

4-39/   Convert $115°$F to the equivalent $°$C.

4-40/   Convert $37°$F to the equivalent $°$C.

4-41/   Copper conductors with Type T insulation are being used to supply a 60 A load, the conductors are in a metallic raceway and the load is located 250 feet from a single-phase 115 volt, 60 Hz ac source, and the maximum ambient temperature is $104°$F. What size conductors are required to limit the voltage drop to 2 percent of the applied EMF?

4-42/   What size conductor is required for Problem 4-41 above if the source voltage is 230 V, all other factors being unchanged?

4-43/   What size conductors would be required for Problem 4-41 with a 100 A load if all other factors remain unchanged?

4-44/   Determine the size conductor you would use in Problem 4-41 if consideration were given to ampacity alone. What does this determination emphasize?

5

Circuit Components

5.1 *Switches and Their Use as Control Devices*

In Sec. 1.4 we noted that all practical electric circuits contain four parts: *source, load, conductors,* and one or more *control devices* to turn the circuit on and off under normal conditions. In this section we are concerned with the control devices or, as they are more commonly known, *switches,* and the ways in which they are connected.

SPST switches

The simplest form of switch is designated *single-pole single-throw* (SPST) to indicate that it opens or closes *one* conducting path and is placed in a designated position by a single *throw* or switching movement. Common forms of the SPST switch include *knife, toggle, push-button,* and *pull-chain* switches. All have *two* terminals. The knife switch has an open mechanism; but in the toggle, push-button and pull-chain switches the switching mechanism is enclosed for improved safety and mechanical protection.

Since an SPST switch can either open or close only one conducting path, it must be connected in *series* with the circuit or device to be controlled.

Thus, when the switch is *closed* (*on*), the circuit is completed, current passes from the source, through the switch and the load device or devices and back to the source. When the switch is *open* (*off*), the circuit is no longer continuous, circuit current is zero, and no electrical energy is supplied to the load device or devices.

Should the switching mechanism become defective—and this is true of any type switch—it may leave the circuit either permanently open or permanently closed. In either case, the switch no longer performs its intended function and must either be repaired or replaced. Except for knife switches, replacement is generally the better course of action.

Two methods of connecting an SPST switch are illustrated in the schematics in Figs. 5-1(a) and (b). As noted earlier, the switch, in both cases, is connected in series with the load to be controlled. In Fig. 5-1(a) however, each component of the load is also series-connected, while in Fig. 5-1(b) the load consists of three parallel-connected lamps. The source may, of course, be either ac or dc. Because an ac source is most common in residential wiring, however, this type of source is indicated throughout the remainder of this text except in Chapter 13 which relates to motors.

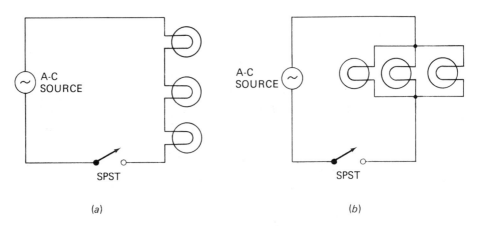

(a)   (b)

Figure 5-1   Wiring arrangements for SPST switches. In both cases, the switch is series connected. The load, however, may consist either of series-connected devices, as in (a), or parallel-connected devices, as in (b).

Notice the symbol for an SPST switch. Also note that in *any schematics* in this text, the physical connection of two or more conductors is always indicated by a heavy dot at the point of intersection between lines representing the conductors. Where lines cross and the heavy dot is absent from the schematic at the point of intersection, there is no physical connection of conductors. In some literature, however, a physical connection is understood to exist wherever two lines intersect (no dot), while the absence of connection is indicated by a bridge-like symbol (⌒).

DPST switches

A *double-pole single-throw* (DPST) switch has two *ganged* (mechanically joined) switching mechanisms which *simultaneously* open or close *two* conducting paths with a *single* throw of the switch. The fact that the switching mechanisms are ganged is indicated schematically by a dashed line. Again, the load may consist of either series- or parallel-connected devices. See Figs. 5-2(a) and (b).

Use is made of DPST switches, instead of the SPST variety, when it is necessary to open two *ungrounded* conductors. (The subject of grounding is taken up in Chapter 6.)

Switches of the DPST type have four terminals and, on toggle switches, the words "on" and "off" appear on the handle (throw lever).

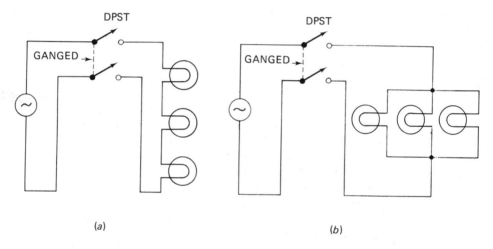

(a)   (b)

Figure 5-2   Wiring arrangements for DPST switches. The two switch mechanisms are mechanically ganged so that both wires are either opened or closed simultaneously. The load may consist of either series-connected devices, as in (a), or parallel-connected devices, as in (b).

SPDT switches

When it is either necessary or desirable to control a load from *two* locations (for example, upstairs and downstairs in a home), two *single-pole double-throw* (SPDT) switches may be employed. A typical arrangement is shown in Fig. 5-3. With both switches in the same position, the circuit is closed and current is supplied to the lamp. If either switch is then thrown to its opposite position, the circuit is open.

As indicated in the Fig. 5-3 schematic, an SPDT switch, also referred to in the electrical trade as a *3-way* switch, has *three* terminals. The *common*

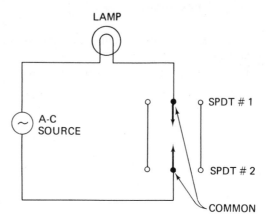

Figure 5-3   Two SPDT switches used to control a lamp from two locations, such as the upstairs and downstairs of a home. Switches of this type are also called 3-way switches.

terminal usually has a dark or oxidized finish. Notice that the common terminal of one switch connects to one end of the load while the common terminal of the other switch connects to one end of the source. The remaining four terminals are connected as *two-terminal pairs*, and the wire between each pair is called a *runner* or sometimes, *traveler*. Mechanical differences in the switches produced by different manufacturers may cause some slight difficulty in connecting the switches properly, but no danger exists if the wrong terminals are chosen initially. The circuit will not work properly, of course, until the correct connections are made.

The words *on* and *off* do not appear on the handle of a 3-way switch.

DPDT switches

To control a load from *three* or more locations, use is made of *double-pole double-throw* (DPDT) switches (also called *4-way* switches) in conjunction with *two* SPDT switches.

As in any switch having more than one switching mechanism, the switching mechanisms are ganged and, therefore, work in unison.

Typical wiring arrangements for controlling a light from *three* locations are shown in Figs. 5-4(a) and (b). Notice that the internal switching action and, therefore, the wiring of the DPDT switch used in (a) differs from that used in (b). Moreover, the common terminal of one SPDT switch connects to one end of the load, the common terminal of the second SPDT switch connects to one end of the source, and the DPDT switch is inserted between the two SPDT switches.

In Fig. 5-4(a) the light is *off*. When any one of the three switches is

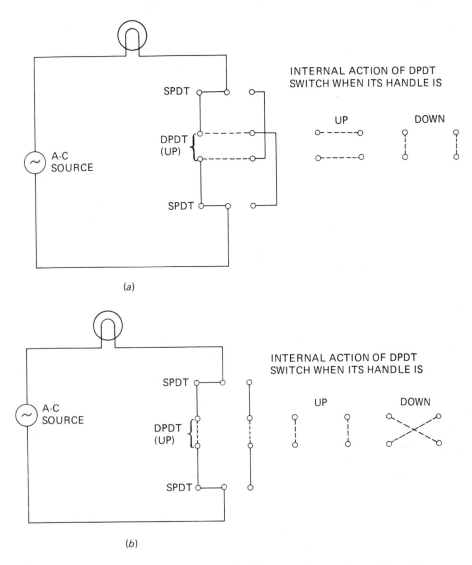

Figure 5-4   Controlling a light from three locations. The different internal switching actions of the DPDT switches necessitates the different wiring arrangements shown in (a) and (b).

thrown to its opposite position, however, the circuit is completed and the light turns *on*.

In Fig. 5-4(b) and light is *on* and moving the lever of any one of the three switches turns the light *off*.

Examination of either circuit with the switches in any combination of positions will show that the desired control is, indeed, achieved.

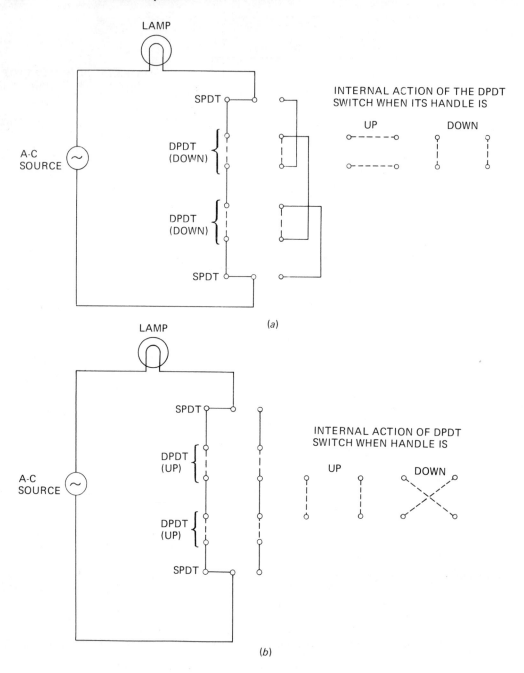

(a)

(b)

Figure 5-5   Controlling a load from four locations. The different internal switching actions of the DPDT switches necessitates the different wiring arrangements shown in (a) and (b). Schematic symbols for (a) thermal and (b) magnetic circuit breakers.

When load control from more than three locations is required, two SPDT switches are again used with their common terminals connected to one side of the load and source, respectively; and the required number of DPDT switches is connected between them—one additional DPDT switch for each additional location. Two wiring arrangements for controlling a load from four locations are shown in Figs. 5-5(a) and 5-5(b).

Checking the internal action of DPDT switches

The internal action of DPDT switches can be checked quickly and easily by means of a *continuity checker*. Although many different forms of such testers are available, my own preference is the ohmmeter section of a combination volt-ohmmeter. A wide variety of inexpensive instruments of this type are currently available. All are small in size, lightweight, and completely safe in operation.

As indicated in Figs. 5-4 and 5-5, a DPDT switch has four terminals. On the one hand, if continuity (essentially the zero reading on an ohmmeter) is indicated when the test prods of the tester are first placed across the upper terminals and then across two terminals located on the same side of the switch, the switch has the internal action shown in Fig. 5-5(a). If, on the other hand, continuity is indicated when the test probes of the tester are first placed on the two terminals located on one side of the switch and then on terminals located at diagonal corners, the switch has the internal action shown in Fig. 5-5(b).

With either type of switch, of course, physical movement of the switch handle is required to obtain the correct readings.

5.2   *Basic Code Provisions on Switches*

Article 380 of the Code relates to switches and their application. In Article 100 a *general-use switch* is defined as one intended for use in general distribution and branch circuits. It is rated in amperes and is capable of interrupting the rated current at its rated voltage.

A *general-use snap switch* is defined as a form of general-use switch so constructed that it can be installed in flush device boxes or on outlet box covers used in conjunction with wiring systems recognized by the Code. (Wiring methods are covered in Chapter 7.)

Two types of general-use snap switches are considered in Sec. 380-14; AC and AC-DC. The AC type is readily identified by the letters AC stamped on the switch strap. The absence of any such designation indicates an AC-DC switch.

All switches are marked with the rated current and voltage and, if horsepower rated, the maximum rating for which they are designed.

AC switches

Switches of this type are suitable only for use on ac circuits for controlling the following: (1) resistive and inductive loads, including electric-discharge lamps such as the fluorescent type, not exceeding the ampere rating of the switch at the voltage applied; (2) tungsten filament lamp loads not exceeding the ampere rating of the switch at the voltage applied; and (3) motor loads not exceeding 80 percent of the ampere rating of the switch at its rated voltage.

AC-DC switches

Swtiches of this type are suitable for use on either AC or DC circuits for controlling the following: (1) resistive loads not exceeding the ampere rating of the switch at the voltage applied; (2) inductive loads not exceeding 50 percent of the ampere rating of the switch at the applied voltage. Switches rated in horsepower are suitable for controlling motor loads within their rating at the voltage applied; and (3) tungsten filament lamp loads not exceeding the ampere rating of the switch at the applied voltage, when "T" rated.

T-rated switches

As noted in Sec. 3.2 of this text, the tungsten filament is the principal light source in incandescent lamps. The *cold* resistance of a tungsten filament is very small compared to its *hot* resistance. As a result, a very large current flows when the light is first turned on.

The high *inrush* current, although short in duration (less than 0.2 s), necessitates heavy contacts on the switches used in incandescent lighting circuits. Switches of this type are called *T-rated* switches, and their listed current rating is for the lamp's transient inrush current. A 100 W tungsten filament, for example, draws a normal (hot) current of 0.87 A at 115 V, but its inrush or starting current is about 9 A. Such a lamp requires a switch with a 10 A T-rating. When several incandescent lamps are connected in parallel and are controlled by a single switch, the T-rating of the switch should exceed the total inrush-current requirement.

The accessibility, grouping, and mounting of general-use snap switches is considered in Chapter 10.

Today, the AC type switch is used in practically all new wiring and as replacement for the older AC-DC type. The widespread distribution of ac by electric utility companies, as well as the lower cost, and quieter operation of AC snap switches account for their popularity.

Switch miscellany

An *isolating* switch is one intended to isolate an electric circuit from the source of power. It has no interrupting rating, and is intended to be operated only after the circuit has been opened by some other means.

A *motor-circuit* switch is one rated in horsepower, capable of interrupting the maximum operating overload current of a motor of the same horsepower rating as the switch at the rated voltage.

Surface-type switches used with open wiring on insulators must be mounted on sub-bases of insulating material which separate the conductors at least one-half inch from the surface wired over.

A switch or circuit breaker in a wet location or outside a building must be enclosed in a weatherproof enclosure or cabinet placed in such a way as to prevent the entrance and accumulation of moisture or water. At least one-quarter inch of air space is required between the enclosure and the supporting surface.

5.3   *Receptacles and Face Plates*

In Article 100 a *receptacle* is defined as a contact device installed at an outlet for the connection of an attachment plug. The outlet itself is a point on the wiring system at which current is taken to supply a load.

Today, most receptacles are of the duplex type and accommodate two plugs. When purchased, the two receptacles are connected in parallel. In better quality receptacles, however, a small metal strip can be broken out to permit isolation of the two sections. With this arrangement, the top receptacle can be wired so that it is permanently *live* and convenient for supplying a temporary load, such as a vacuum cleaner. The bottom half can then be used for lamps and is switch controlled.

Receptacles installed for the attachment of portable cords must be rated at not less than 15 A, 125 V, or 10 A, 250 V.

For safety reasons, switches and receptacles are covered with *face plates* after installation.

Face plates may be either metallic or nonmetallic. Cutouts are provided for various switch and/or receptacle combinations, and the color can be matched to the room decor.

5.4   *Lamp Holders [Sockets]*

Probably the most common form of *lamp holder* is found in residential wiring. Usually referred to as a *socket* it is the brass-shell type and ordinarily contains either a key, a push-through, or a pull-chain switching mechanism to control the lamp. The metal shell and cap are lined with insulating material to keep them from becoming part of the electrical circuit (Sec. 410-50).

Lamp holders with a porcelain shell and pull-chain switching mechanism are often used on top of outlet boxes. A weatherproof type is available for outdoor use.

5.5   *Devices for Overcurrent Protection*

Any electrical system must include safeguards to assure that the ampacity of each conductor is not exceeded. An excess current, called *overcurrent* (or, sometimes, *fault current*), may range from a small overload to a complete short circuit, depending on the location of the circuit fault.

Since the electric power producing heat is equal to $I^2 R$, a given current through a conductor which is double its ampacity has heat in that conductor four times its normal value. When short-circuit conditions exist, even for a few seconds, the conductor melts and the insulation ignites and sets fire to any surrounding materials which are not fireproof. Overcurrent protection assures that the current will be interrupted before excessive current causes damage, however, either to the conductors themselves or the load they supply.

There are basically two classes of devices used in residential wiring for overcurrent protection: *fuses* and *circuit breakers*. Neither device can be arranged or installed in parallel, except those circuit breakers which are assembled in parallel and tested and approved as a single unit. Moreover, no overcurrent-protection device can be placed in a permanently grounded conductor—except where it simultaneously opens all conductors of the circuit. In residences, overcurrent devices must be located where they are readily accessible, not exposed to physical damage, and not in the vicinity of easily ignitible materials.

Fuses—general

Fuses are overcurrent devices that destroy themselves when they interrupt the circuit. They are made of a low-melting temperature metal and calibrated so that they melt when a specific current rating is reached. Because such fuses are connected in series with the load, they open the circuit when they melt. All fuses have an *inverse time* characteristic. Thus, a fuse rated at 30 A should carry 30 A continuously; with about a 10 percent overload it should melt in a few minutes and with a 20 percent overload in less than a minute. A 100 percent overload should cause a correctly-rated fuse to blow or melt in a fraction of a second.

Plug fuses (Sec. 240-E)

In plug fuses of the Edison-base type, a fusible link is enclosed in a housing which prevents the splatter of molten metal when the fuse blows. The condition of the fuse is determined by looking through a window located on top of the plug assembly. If the window (or other prominent feature) is *hexagonal* in shape, the fuse is rated at 15 A or less. If the window is *round*, the fuse is rated at more than 15 A, up to a *maximum* size for fuse plugs of 30 A.

Code Sec. 240-50(a) states that plug fuses and fuse holders shall not be used in circuits exceeding 125 volts between conductors, except in circuits supplied from a system having a grounded neutral, and that no conductor in such circuits is to operate at more than 150 volts to ground. Thus, the use of plug fuses *is permissible* between the two hot wires of a 115/230 volt installation with a grounded neutral since the voltage from either conductor to ground is 115 volts.

Type S fuses (Sec. 240-54)

Type S fuses and fuse holders were designed specifically to prevent either accidental or intentional replacement of a given fuse with one of a higher rating.

When a Type-S fuse is used with an Edison-type fuse holder, a non-removable adapter is screwed into it and locked in place. When a Type-S fuse is inserted in the adapter it should be screwed in tighter than usual to ensure that the spring under the shoulder of the fuse flattens.

Type-S plug fuses and fuse holders are classified at not over 125 V; 0 to 15 A, 16 to 20 A, and 21 to 30 A. Fuses of the 16 to 20 A and 21 to 30 A classifications will not fit fuse holders or adapters of a lower ampere classification.

The fuses, fuse holders, and adapters from all United States manufacturers are interchangeable.

Cartridge fuses (Sec. 240-F)

Whenever the circuit current exceeds 30 A, it is necessary to use a cartridge-type fuse and fuse holder.

The voltage and current ratings of cartridge fuses and fuse holders are shown in Table 5-1.

Fuse holders must be designed so that is is difficult for one to put a fuse of given class into a fuse holder designed for a lower current or higher voltage than that of the class to which it belongs.

Fuses must also be plainly marked to show ampere rating, voltage rating, interrupting rating where other than 10,000 A, *current limiting* where

Not over 250 V amperes	Not over 300 V amperes	Not over 600 V amperes
0-30	0-15	0-30
31-60	16-20	31-60
61-100	21-30	61-100
101-200	31-60	101-200
201-400		201-400
401-600		401-600

Table 5-1   Current and voltage classifications for 0—600 ampere cartridge fuses and fuseholders

applicable, and the name or trademark of the manufacturer. (Exception: The interrupting rating may be omitted on fuses used for supplementary protection; that is, fuses used with appliances or other equipment to provide individual protection for specific components or circuits within the equipment itself.)

The current-limiting fuse referred to above is a device which, when interrupting a specific circuit, consistently limits the short-circuit current to a value substantially less than that which would exist if a solid conductor of comparable resistance were used in place of the fuse.

Current-limiting fuses cannot be installed in noncurrent-limiting fuse holders.

Circuit breakers (Sec. 240-G)

Article 100 defines a circuit breaker as a device designed to open and close a circuit by non-automatic means, and to open the circuit automatically on a predetermined overload of current, without injury to itself when properly applied within its rating.

To energize a circuit breaker, the *on-off* switch or button is thrown to the "on" position. This closes the contacts of the circuit breaker which are held closed by a latch. When the *on-off* switch is turned "off" the latch is released and the contacts are reopened by spring action.

Some circuit breakers have a *thermal element* connected in series with the contacts and are called *thermal* circuit breakers. Their function is to protect against *gradual* overload conditions. The current passes through the series-connected thermal element and causes it to heat. When excessive heat is produced, as a result of increasing overload condition, a *bimetallic strip* bends and releases the contact-holding latch automatically. The bimetallic strip is made by having thin strips of two different metals in contact with one another.

Because it takes time for a bimetallic strip to heat, the tripping or opening of a thermal circuit breaker *does not* occur the instant a current exceeds the rated value of the circuit breaker. The manufacturer supplies an operating characteristic in graphic form with each circuit breaker. This curve shows how long it takes for the breaker to trip for a given current. Whenever you select a thermal circuit breaker for a given application, do so by referring to the characteristic curve. Needless to say, a thermal circuit breaker should never be used in any application where a very fast response to overload conditions is required.

Another type of circuit breaker contains an *electromagnet* which is connected in series with the circuit to be protected. Normal circuit current has no effect on the electromagnet; however, when a short circuit occurs, the electromagnet trips the breaker immediately. Thus, a magnetic circuit breaker provides *instantaneous* protection against a short circuit.

Combination thermal-magnetic circuit breakers are also available to protect against both gradual overload and short circuits. Such circuit breakers can be adjusted to open instantaneously for a short-circuit condition and to open at a lower abnormal current after a time delay.

Location of overcurrent-protection devices (Sec. 240-21)

In general, an overcurrent-protection device is required at the point where each conductor receives its supply.

Eight exceptions to the general rule are listed in the Code. To avoid confusion, however, these exceptions will be treated where applicable.

5.6   *Outlet, Junction, and Switch Boxes*

For safety reasons, appropriate boxes or receptacles (Article 370) must be used at every point where conductors are joined or connected to the terminal of outlets or switches.

In shape, these boxes may be octagonal, square, or round. Their materials may be metallic or nonmetallic and according to their depth, either *deep* or *shallow*. Any box less than 1-1/2″ deep is considered a shallow box. Easily removed knockouts are provided to make holes for the insertion of conductors, but any holes not used must be closed.

Round boxes cannot be used where conduits or conductors requiring the use of locknuts or bushings are connected to the side of the box.

Nonmetallic boxes are generally used with nonmetallic sheathed cable and with approved nonmetallic conduit.

Metallic boxes used with nonmetallic sheathed cable and in contact with any metal surface must either be insulated from their supports and from the metal surface or they must be grounded.

Weatherproof boxes and fittings are required for use in wet locations.

Specially-designed boxes are required for floor-mounted receptacles and for lighting-fixture outlets. Most often the boxes are supported from a structural member of the building either directly or by using an approved metallic hanger or a wooden brace not less than 1″ thick.

In new walls where no structural members are provided, or in existing walls in previously wired buildings, boxes not over 100 cubic inches in size, specifically approved for the purpose, can be affixed with approved anchors or clamps.

Threaded boxes or fittings not over 100 cubic inches in size which do not contain devices or support fixtures can be supported by two or more conduits threaded into the box wrench tight and supported within 3′0″ of the box on two or more sides.

Outlet boxes for concealed work must have an internal depth of at least 1-1/2″, except that where the installation of such a box will result in injury to the building structure or is impracticable, a box of not less than 1/2″ internal depth may be installed.

In completed installations each outlet box must have either a cover or a fixture canopy. All pull boxes, junction boxes and fittings must have an approved cover.

Covers of outlet boxes having holes through which flexible cord pendants pass, must have either bushings designed for the purpose or smooth, well-rounded surfaces on which the cords may bear. Hard rubber or composition bushings cannot be used.

Number of conductors in a box (Sec. 370-6)

Boxes must be of sufficient size to provide free space for all conductors enclosed in the box. Before noting Code limitations, let us define three terms.

Fitting — An accessory such as a locknut, bushing, or other part of a wiring system which is intended primarily to perform a mechanical rather than an electrical function.

Device — A unit of an electrical system which is intended to carry but not utilize electric energy. In this section of the Code, the word device is taken to mean a switch box.

Hickey — A threaded coupling for attaching a fixture to an outlet box.

Fixture wires excepted, the maximum numbers of conductors permitted in outlet and junction boxes must be as in Table 5-2.

Table 5-2 applies where no fittings or devices such as fixture studs, cable clamps, hickeys, switches or receptacles are contained in the box and where no grounding conductors are part of the wiring within the box. Where one or more fixture studs, cable clamps, or hickeys are contained in the box, the number of conductors must be one less than shown in the tables; an additional reduction of one conductor must be made for each strap containing one or more devices; and a further reduction of one conductor must be made for one or more ground conductors entering the box. A conductor running through the box is counted as one conductor, and each conductor originating outside the box and

Box dimensions in inches (trade sizes)	Cubic inch capacity	Maximum number of conductors			
		No. 14	No. 12	No. 10	No. 8
4 × 1-1/2 Octagonal	15.5	7	6	6	5
4 × 2-1/8 Octagonal	21.5	10	9	8	7
4 × 1-1/2 Square	21.0	10	9	8	7
4 × 2-1/8 Square	30.3	15	13	12	10
4-11/16 × 1-1/2 Square	29.5	14	13	11	9
4-11/16 × 2-1/8 Square	42.0	21	18	16	14
3 × 2 × 1-1/2 Device	7.5	3	3	3	2
3 × 2 × 2 Device	10.0	5	4	4	3
3 × 2 × 2-1/4 Device	10.5	5	4	4	3
3 × 2 × 2-1/2 Device	12.5	6	5	5	4
3 × 2 × 2-3/4 Device	14.0	7	6	5	4
3 × 2 × 3-1/2 Device	18.0	9	8	7	6
4 × 2-1/8 × 1-1/2 Device	10.3	5	4	4	3
4 × 2-1/8 × 1-7/8 Device	13.0	6	5	5	4
4 × 2-1/8 × 2-1/8 Device	14.5	7	6	5	5

Table 5-2   Boxes

terminating inside the box is counted as one conductor. If no part of a conductor leaves the box, it is not counted. For example, a wire from the green terminal of a receptacle to a grounding point on the box is not so "counted." The volume of a box is the total volume of all assembled sections.

For combinations or conductor sizes not shown in Table 5-2, those in Table 5-3 apply.

Size of conductor	Free space in cubic inches within box for each conductor
No. 14	2.00
No. 12	2.25
No. 10	2.50
No. 8	3.00
No. 6	5.00

Table 5-3   Volume required per conductor

Exceptions to the above noted Code specifications will be dealt with as the need arises.

EXERCISES

5-1/   With respect to the load, how must a switch be connected in any circuit? Why?

5-2/   When is a DPST switch used in preference to the SPST type?

5-3/   Which types of switches would you select to control a load from five locations? How would you connect them? Draw the circuit.

5-4/   What types of general-use snap switches are available? In what types of circuits may each type be used?

5-5/   Which type of general-use snap switch is most often used in practice? Why?

5-6/   What is a *T-rated* switch?

5-7/   With what types of load must a T-rated switch be used?

5-8/   What is an *isolating* switch?

5-9/   What is a *motor-circuit* switch?

5-10/   What is a *surface*-type switch?

5-11/   What is meant by the statement that all fuses have an *inverse time* characteristic?

5-12/   What is the maximum permissible ampere rating of a plug fuse?

5-13/   How are plug fuses rated at 15 A or less identified?

5-14/   How are plug fuses rated at more than 15 A identified?

5-15/   Does the Code permit the use of plug fuses between the two hot wires of a 115/230 volt installation? Explain fully.

5-16/   How do Type S plug fuses differ from the Edison-base type?

5-17/   When must a *cartridge*-type fuse be used?

5-18/   What is a *current-limiting* fuse?

5-19/   How does a *thermal*-type circuit breaker operate?

5-20/   How does a *magnetic*-type circuit breaker operate?

5-21/   For protection against gradual overload, what type of circuit breaker would you select? Why?

5-22/   What Code restriction exists on the use of round boxes?

5-23/   What is a *shallow* box?

5-24/   When does the Code permit the use of a shallow box?

5-25/   How is the permissible number of conductors in a box determined?

Part
3

6

System and Equipment Grounding

6.1 Grounding Methods

The words "ground" and "earth" mean the same thing. Thus, when any part of an electrical system is grounded it is connected, either directly or indirectly, to earth.

For reasons of safety, the *neutral* wire of all residential-wiring systems is intentionally grounded. Why this is necessary is examined in detail in Sec. 6.2. Such grounding is illustrated in Fig. 6-1 which shows a typical overhead-type residential-supply installation.

A *service drop*, supplied and installed by the utility company, is made up of three *service conductors* which extend from the street main or transformer to the building being served. Notice that the service conductors terminate at insulators mounted on the side of the building. The neutral service conductor is grounded at the distribution transformer.

Three *service-entrance conductors* attach to the service conductors near their point of termination. The service-entrance conductors are often made up either in the form of a cable or are, as in Fig. 6-1, enclosed in conduit. Either form of conductor enclosure is called the *service raceway*.

The service-entrance conductors first connect to a meter which records the number of watthours of electric power consumed. From the meter these

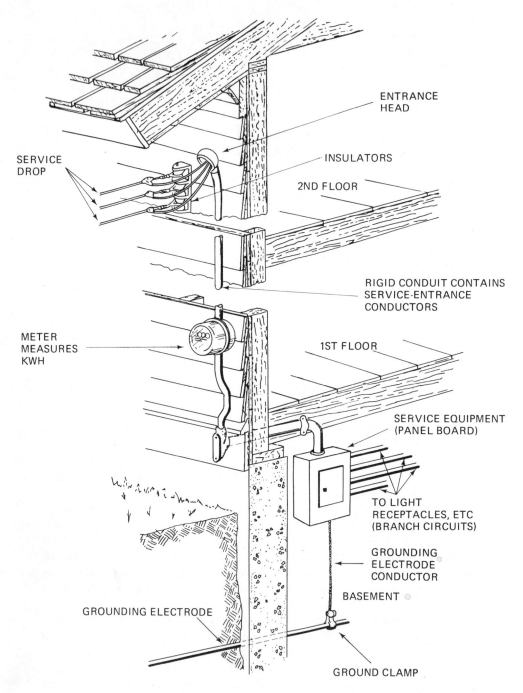

ENTRANCE
HEAD

SERVICE
DROP

INSULATORS

2ND FLOOR

RIGID CONDUIT CONTAINS
SERVICE-ENTRANCE
CONDUCTORS

METER
MEASURES
KWH

1ST FLOOR

SERVICE EQUIPMENT
(PANEL BOARD)

TO LIGHT
RECEPTACLES, ETC
(BRANCH CIRCUITS)

GROUNDING
ELECTRODE
CONDUCTOR

BASEMENT

GROUNDING ELECTRODE

GROUND CLAMP

Figure 6-1   A typical electric-service entrance

conductors extend through the wall of the building and connect to a service switch and fuses (or to a circuit breaker) and accessories which constitute both the main control and a means for cutoff of the electric supply. Collectively, these items are called the *service equipment* and, for safety reasons, they are mounted in a metal enclosure.

At the point where the neutral wire of the service-entrance conductors attaches to the service switch, a *grounding-electrode conductor* is connected. The opposite end of this conductor attaches to the *grounding electrode*. In Fig. 6-1 the grounding electrode is an underground metal water pipe. Where available, a metal water pipe, regardless of its length, and whether it is supplied by a community or local underground water piping system or by a well on the premises, is *always* used as the grounding electrode.

Where a water system like that described above is not available, the grounding connection may be made to the metal frame of a building which is effectively grounded, or to a continuous metallic gas-piping system, or to any other metallic underground systems—such as piping, tanks, and the like. The grounding connection can also be made to the concrete-encased steel reinforcing bar or rod systems in underground footings or foundations or to a *made* electrode constructed from a pipe, rod, plate or other approved device driven into the ground. (See Code Sec. 250-H for complete information on acceptable grounding electrodes.)

Generally, the service drop, service-entrance conductors, watthour meter, service switch and fuses (or circuit breakers), and the ground are referred to collectively as the *service entrance*.

Physically, overcurrent protection devices for *branch* circuits—that is, circuits which supply lighting and appliance outlets—are most always located in the same metal enclosure as the service equipment. For that reason, they are usually considered to be part of the service entrance.

The neutral wires of all branch circuits are connected to the grounded neutral wire of the service-entrance conductors. In this way the neutral wire of all residential branch circuits is grounded. The grounded neutral wire of each branch circuit must be insulated and, most often, the insulation is *white* in color. Although the NEC also permits the insulation of the grounded neutral conductor to be natural gray in color, we shall consider it and refer to it as *white* throughout the remainder of this text.

Two notable exceptions to the specified identification scheme are permitted by the NEC and deserve comment at this point in our discussion.

- Because No. 4 and heavier wires are rarely any color other than black, they may be used for the neutral grounded conductor, *provided* the ends of the black wire are *painted white*.

- Outdoor wire is available only with black insulation. Thus, white wire is not required for the grounded neutral in outdoor installations.

Now, here is a most important fact that you must remember:

In residential wiring, the grounded neutral wire is never interrupted by a fuse, circuit breaker, switch or other device. Without exception, the white neutral wire always runs directly from the point of attachment to the grounding electrode to the outlet from which electric power is finally channeled for consumption.

The grounded neutral service-entrance conductor and the grounded neutral (white) conductor of each branch circuit form what is called the *system* ground of an electrical installation.

In addition to the system ground, the NEC requires that all metallic equipment—raceways, boxes, fittings, enclosures for fuses or circuit breakers, and so forth—associated with an electrical system be grounded. This type of grounding is called *equipment grounding*.

6.2   *Why is Grounding Necessary?*

The principal reason for system grounding is to limit the "voltage to ground" from rising above a safe value because of a fault *outside* the building. If, for example, outside wires are struck by lightning, the ground acts like a lightning rod to get rid of the very high electrical charges built up on the wires. Furthermore, if neutral service-entrance conductors are not grounded, a faulty power-supply transformer could also cause voltages inside the building to rise far beyond their normal values.

Equipment grounding is necessary to prevent electric shock to persons coming into contact with metallic objects which, either intentionally or accidentally, form part of the electrical system.

If, for example, an ungrounded (hot) conductor in the system should accidentally touch any metallic equipment, junction box, motor frame, or fitting, then there is voltage between that equipment and ground. So if that equipment is not grounded, a very serious shock hazard exists.

To illustrate this problem, study Fig. 6-2 which shows an ungrounded raceway connected to a box containing a switch with a wall plate attached to that switch. If the hot (black) conductor of the branch circuit accidentally comes into contact with that switch box, both box and wall plate are at a potential of 115 volts above ground. Should anyone in contact with a grounded object now touch that wall plate, his body would immediately complete the circuit to ground. The resulting shock would be most unpleasant—at best. Depending on the resistance of that person's body and the condition of the ground (dry, damp, or wet), the shock could even prove fatal.

The type of dangerous shock described above is avoided by grounding the raceway and, therefore, the box and wall plate to which it is connected. In this case, if the hot conductor accidentally comes into contact with the box, it is the same as having the black and white conductors come into contact with each other. The resulting short circuit causes the overcurrent protection device of the branch circuit to "blow." The circuit is then "dead" since the blown protective

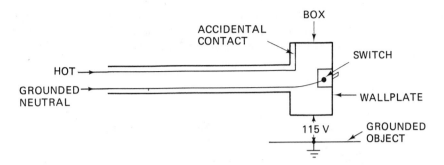

Figure 6-2   As a result of accidental contact between a hot conductor and the switch box, a potential of 115 V exists between the entire raceway system and the ground

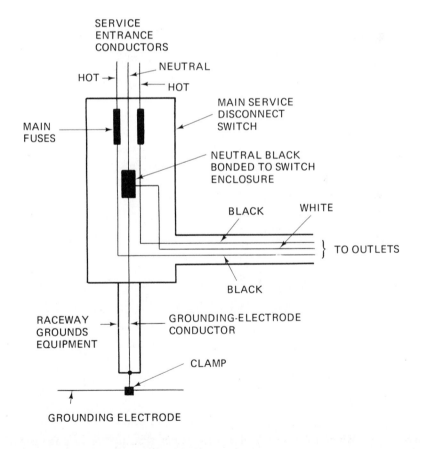

Figure 6-3   System and equipment grounding

device creates an open circuit. The current that causes the protective device to become an open circuit is usually called a *fault current* since it flows when, and only when, there is a fault in the circuit.

The NEC requires that all equipment be grounded to the same electrode (usually a water pipe) that grounds the system. This is shown in Fig. 6-3.

6.3   *Are All Systems Grounded?*

Insofar as residential wiring is concerned, the general answer is "yes": all systems are grounded. Because the Code tends to be rather involved on this point, however, further clarification should prove beneficial.

Sec. 250-5(b)(1) states that ac systems of 50 to 1000 volts supplying premises wiring shall be grounded when the maximum voltage to ground on ungrounded conductors *does not exceed* 150 volts.

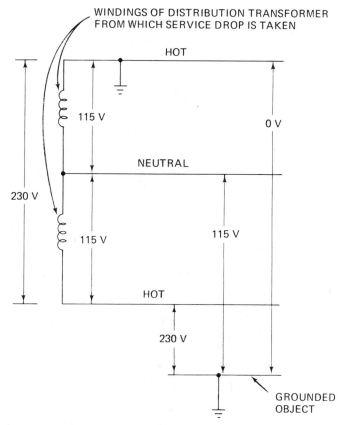

Figure 6-4   When a hot line is grounded, rather than neutral, a potential of 230 V exists between one side of the line and the ground

As you now know, a 115-230 V arrangement is generally used when a single-phase source is required. The choice of which conductor to ground should be obvious. As shown in Fig. 6-4, if we were to ground one of the hot conductors instead of the neutral, the voltage to ground from one side of the line would be 230 V, and this is a *code violation.*

In the case of a 120-208 V, four-wire, three-phase system (Sec. 2.6), the neutral conductor is again grounded. See Fig. 6-5. Since each load then has a grounded conductor, the voltage to ground from any "hot" conductor is 120 V. This meets code requirements.

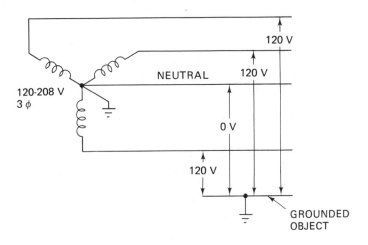

Figure 6-5   Voltage to ground in a 120-208 V three-phase supply with the neutral conductor grounded

Another NEC requirement is that, when single-phase loads are connected from any line wire to ground, each single-phase load has a grounded conductor. All single-phase systems supplying lighting circuits must be grounded.

Now let us consider a 230 V, three-wire, three-phase system used for power rather than for lighting. In this case, the NEC recommends, but does not require, system grounding if the voltage to ground can be less than 300 V. The grounding of any one conductor would fulfill this recommendation. When a question arises in any such case, the decision is made by the power supplier.

To sum up, then, if the voltage to any other grounded object can be kept under 150 V, the system must be grounded. This includes all residential electrical systems and/or any system that uses incandescent lamps.

Any higher-voltage system used to supply power to motors or industrial equipment may or may not be grounded. The decision of local authorities prevails in all such cases.

6.4   *Voltage to Ground*

We have used the expression "voltage to ground" several times without defining it. What exactly does the term mean?

"Voltage to ground" means simply the potential from any point in the system to any other object that itself is grounded.

After a system is grounded, the voltage to ground from any ungrounded conductor must be the same as the voltage to the grounded conductor. Figs.

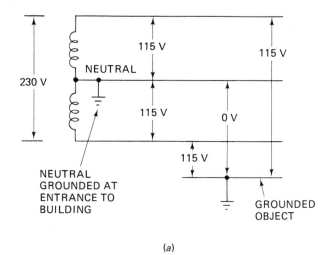

(a)

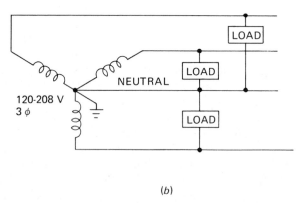

(b)

Figure 6-6   In properly grounded single-phase, three-wire systems (a) and three-phase, four-wire systems (b), the voltage to ground from any ungrounded conductor is the same as the voltage to the grounded neutral

6-6(a) and (b) illustrate this for 115-230 V, single-phase and 120-208 V three-phase systems, respectively.

Theoretically, in the three-phase ungrounded system of Fig. 6-7, there should be no measurable voltage to ground from any conductor since there is no electrical connection between any conductor and ground. In practice, however, the ac in an extensive system produces a so called "capacitive effect" which induces some measurable voltage between the conductors and raceways. Since the raceways are grounded, this induced voltage is a voltage to ground.

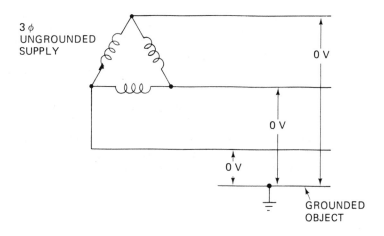

Figure 6-7   Voltage to ground in an ungrounded system

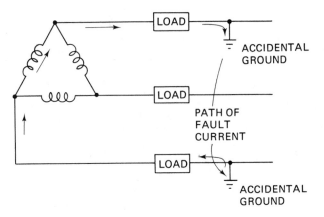

Figure 6-8   Fault current when two accidental grounds exist in an ungrounded system

In complicated electrical systems unintentional grounds may occur, such as that resulting from moisture in a motor winding. In such a case, the system would behave as though it were intentionally grounded and the voltage to ground might be the same value as the voltage between conductors. For this reason, the NEC defines the voltage to ground in an ungrounded system as the *maximum* voltage of the system.

In Fig. 6-8, two accidental grounds are assumed to occur on two different conductors of a three-phase ungrounded system. If the resistance of the two grounds is low enough, excess current will flow and cause the overcurrent-protection device to open the circuit. If only one accidental ground exists, however, the system will still function normally.

6.5   *More on Shock Hazards and Their Elimination*

As noted earlier in Sec. 6.3, with a 115-230 V, three-wire supply, all lighting systems must have one circuit of the conductor grounded. Why? As indicated in Fig. 6-9, if the screw-shell terminal of an incandescent lamp is incorrectly connected to an ungrounded conductor, a potential of 115 volts exists between the shell and any grounded object. Now, suppose a person changing the lamp simultaneously contacts a grounded object and the shell. His body is then the path by which current returns to ground. The resultant shock he receives may even prove fatal.

It should be obvious from the above discussion that the screw shell of any incandescent lamp should be connected to the grounded conductor. Moreover, the grounded conductor can only be grounded at the point where the service-entrance conductors enter the building. Any additional ground violates the NEC and creates hazards which will be discussed shortly.

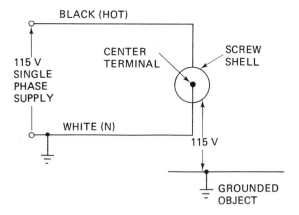

Figure 6-9   When the screw shell is connected to the ungrounded conductor, a potential of 115 V exists between the screw shell and any grounded object and a definite shock hazard exists

The grounded terminal of a lampholder is identified by a nickel- or zinc-colored screw. If a lampholder has wire leads instead of screw terminals the shell is identified by a white or a natural-gray conductor.

In Sec. 6.1 we learned that the grounded neutral in a residential wiring system should never be opened or interrupted by a fuse, circuit breaker, switch or other device. Let us see what happens if this rule is violated.

Refer to Fig. 6-10(a) which shows a single-pole switch connected in series with the grounded conductor. Notice that even with the switch open, a 115 V potential still exists between the screw shell of a lamp holder and ground. The same situation would hold true, of course, for an open circuit breaker or a blown fuse. Anyone in simultaneous contact with the screw shell and any grounded object would receive a stiff "jolt."

The correct method of installing a single-pole switch in series with the "hot" conductor is shown in Fig. 6-10(b). Installed this way, no continuous path

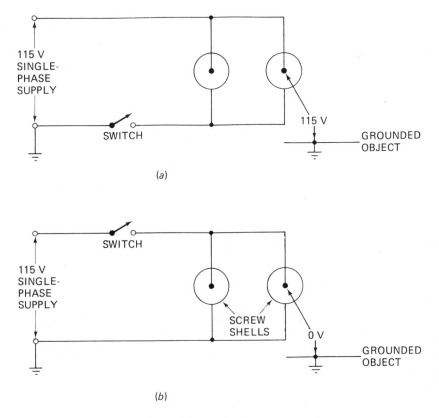

Figure 6-10   When an open single-pole switch is connected in series with the grounded neutral conductor, (a) a potential of 115 V exists between the center terminals of the screw shells and any grounded object. In (b) this hazard is removed by placing the switch in the hot side of the circuit

to ground can possibly exist between the screw and ground when the switch is open. Therefore, there is no possibility of an electric shock.

In some installations other than residential you will find circuit breakers which have a sufficient number of poles to disconnect all conductors (both "hot" and neutral) simultaneously. This may lead to the erroneous conclusion that you can (or should) use fuses in both the "hot" and grounded neutral conductors of a residential wiring system. Do *not*. The hazard of any such double fusing is shown in Fig. 6-11. If the fuse in the hot lead "blows," all well and good. If the fuse in the ground lead opens, however, while that in the hot leads does not, a potential difference of 115 V again exists between the screw shell and any grounded object; it is dangerous!

In summary, keep this rule in mind. Always use a "solid" unbroken neutral with any three-wire, single-phase supply. The same condition also applies to a three-phase, four-wire system. Based on the facts that we have analyzed up to this point, review the proofs that support the validity of this last statement.

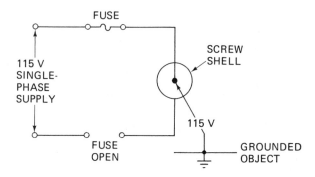

Figure 6-11   Danger of having fuses connected in both sides of a single-phase circuit with the neutral grounded

6.6   *More on Equipment Grounding*

As noted earlier, the grounding of all metal that may contact an ungrounded conductor, either intentionally or unintentionally, is known as equipment grounding; and it is essential to prevent metal equipment from maintaining a voltage above ground (zero potential). In Sec. 6-2 we discussed what could happen with an ungrounded raceway connected to a box containing a switch. Fig. 6-12 depicts a similar arrangement except that the raceway ground equipment and the system ground are connected. Should an accidental ground occur on the ungrounded conductor, the resulting fault current follows the path indicated by the arrows.

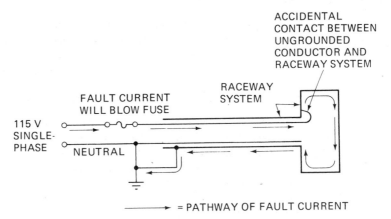

Figure 6-12   When the raceway system is connected to the system ground, fault current will "blow" the fuse and remove any shock hazard

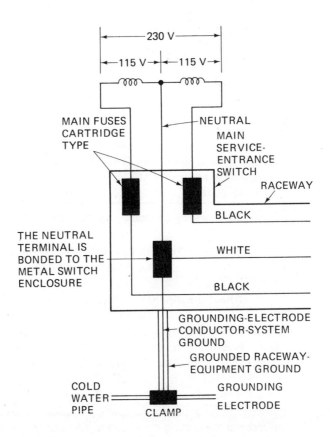

Figure 6-13   System and equipment grounding terminals in service-entrance switch

To permit overcurrent devices to open the circuit, the resistance of the equipment-ground path must be low. Since the resistance of a conductor decreases as its cross-sectional area increases, the use of "thick" conductors ensures the required low resistance. This method of providing a low-resistance ground can be expensive, however, so that construction economics often mean that the use of such conductors must be somewhat limited.

As noted earlier, the system and equipment grounds are joined at the main disconnect (service) switch. Fig. 6-13 shows how this is done. Notice that the grounding-electrode conductor is encased in a metallic raceway and that raceway and conductor are attached to the cold-water pipe by means of a heavy clamp. The equipment ground however, connects only to the raceway of the grounding-electrode conductor, so that there the raceway has a dual grounding function.

In order to ground a wiring system, calculate the size of the grounding-electrode conductor you need from the data in Table 6-1. In those cases where you use made-electrodes, however, the grounding electrode conductor need not be larger than No. 6 copper or its equivalent in ampacity.

Size of service conductor		Size of grounding conductor	
Copper	Aluminum	Copper	Aluminum
2 or smaller	0 or smaller	8	6
1 or 0	00 or 000	6	4
00 or 000	0000 or 250 MCM	4	2
Over 000 to 350 MCM	Over 250 MCM to 500 MCM	2	0
Over 350 MCM to 600 MCM	Over 500 MCM to 900 MCM	0	000
Over 600 MCM to 1100 MCM	Over 900 MCM to 1750 MCM	00	0000
Over 1100 MCM	Over 1750 MCM	000	250 MCM

Table 6-1   Service and grounding conductors for grounded systems

When a service-entrance conductor (system) is not grounded, see the specifications for a proper equipment-grounding conductor (or raceway) in Table 6-2.

When the equipment ground is either conduit or EMT, the low-resistance path that they form permits fault currents to operate overcurrent protective devices easily so the raceways are maintained at ground potential.

In addition to raceways, every metallic part which is not necessary for the operation of an electrical load must be securely grounded. Thus, all fittings and locknuts for boxes, switches, faceplate, motors, and the like must be tight.

Armored cable itself provides a path for fault currents. To reduce the resistance of the ground-return path, a continuous uninsulated aluminum ribbon lies between the paper-covered conductors and the armor. The parallel resistance of the armor and the ribbon is, of course, lower than the resistance of the armor

Size of service conductor		Size of grounding conductor			
Copper	**Aluminum**	**Copper**	**Aluminum**	**Conduit**	**EMT**
2 or smaller	0 or smaller	8	6	1/2	1/2
1 or 0	00 or 000	6	4	1/2	1
00 or 000	0000 ot 250 MCM	4	2	3/4	1-1/4
Over 000 to 350	Over 250 MCM to 500 MCM	2	0	3/4	1-1/4
Over 350 MCM to 600 MCM	Over 500 MCM to 900 MCM	0	000	1	2
Over 600 MCM to 1100 MCM	Over 900 MCM to 1750 MCM	00	0000	1	2

Table 6-2   Service equipment grounding conductor for ungrounded systems

itself. Armored cable must be securely fastened to outlets with connectors or clamps and must not be subject to any strain which would loosen the fastening over a period of time.

In nonmetallic cable an extra conductor is provided for equipment grounding. This conductor may be bare, but if it is insulated, the color of the insulation must be *green*.

Portable appliances, such as floor lamps, which do not use grounding-type receptacles and plugs are always a shock hazard. If the screw-shell insulation breaks down and contacts a metal frame, anyone making simultaneous contact with the frame and any grounded object will be shocked.

Grounding by use of a three-wire cord is illustrated in Fig. 6-14. The receptacle includes a separate grounding prong. The grounding of the appliance is completed through the receptacle to its box and then through the raceway back to the source.

In the special case of major household appliances—dishwashers, electric

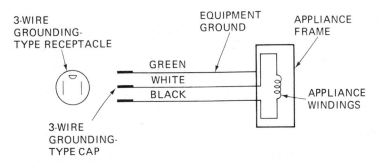

Figure 6-14   Resistance reduces the fault current and the main fuse does not blow. Both shock and fire hazards exist.

ranges, garbage disposals, and the like—the appliance frame is connected to the neutral conductor. This grounding arrangement is not permitted with smaller appliances.

6.7   *More on Fault Currents*

A loose equipment-ground connection prevents a fault current from doing its intended job of activating overcurrent devices. To learn why, study Fig. 6-15 where a poor ground connection between a raceway and box establishes an equivalent resistance of 2.5 ohms. The three-wire, 115-230 V supply is fused at 50 A. A fault current is produced when a "hot" connector comes into contact with the box. The maximum current that can pass through the poor (high resistance) ground connection is, by Ohm's law,

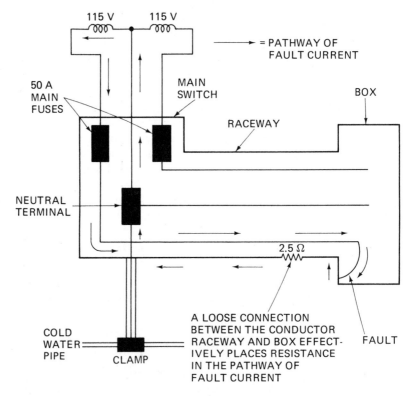

Fig. 6-15   A loose equipment-ground connection prevents a fault current from activating over current devices

$$I = \frac{E}{R} = \frac{115}{2.5} = 46 \text{ A}$$

Under these conditions the 50 A fuse will not blow. As a result, a shock hazard exists between the poor connection and ground. A fire hazard also exists if combustible materials are nearby since any electrical arcing at the poor connection causes flashover and this, in turn, produces unsafe heat.

EXERCISES

6-1/   Explain what is meant by the following terms: **(a)** service drop, **(b)** service conductors, **(c)** service entrance conductors, **(d)** service raceway, **(e)** service equipment, **(f)** grounding-electrode conductor, **(g)** grounding electrode, and **(h)** made-electrode.

6-2/   What colors are permitted for the grounded neutral wire in a residential wiring system?

6-3/   What exceptions, if any, does the NEC permit for the color scheme noted in Exercise 6-2?

6-4/   What is the system ground in a residential wiring system? What is the equipment ground in a residential wiring system?

6-5/   In residential wiring, why must the grounded neutral wire be uninterrupted?

6-6/   Why is a system ground important?

6-7/   Why is an equipment ground important?

6-8/   In a residential wiring system, what danger is inherent in double-fusing; that is, fusing in both the "hot" and neutral lines?

6-9/   Does the NEC permit separate system and equipment grounds?

6-10/   If available, what grounding electrode is always used in residential wiring? Why?

6-11/   Since there are three service-entrance conductors, why does the NEC require grounding of the neutral line?

6-12/   What would happen if, in a residential wiring system, we were to ground a "hot" conductor rather than the neutral conductor?

6-13/   In a 120-208 V, four-wire, three-phase system is any conductor grounded? If so, which conductor and why?

6-14/   Under what conditions must a system always be grounded?

6-15/   Precisely what is meant by the expression "voltage to ground"?

6-16/   How is the grounded terminal of a lamp holder identified?

6-17/   If a lamp holder has wire leads instead of screw terminals, how is the shell identified?

6-18/   In a grounded system using #2 copper service conductors, what size of copper grounding conductor is needed?

6-19/   In an ungrounded system using AWG #0 copper service conductors, what size grounding conductor is needed for **(a)** copper, **(b)** conduit, and **(c)** EMT?

6-20/   What step is taken during manufacture to reduce the resistance of the ground-return path of armored cable?

6-21/   If an insulated grounding conductor is included in nonmetallic cable, what is the required color of the insulation on the grounding wire?

6-22/   Explain how portable-appliance grounding is accomplished with a three-wire cord.

6-23/   How is an electric range grounded?

6-24/   Explain how a loose equipment-ground connection may prevent a fault current from activating an overcurrent device.

6-25/   If an ungrounded cable contacts a cold-water pipe intermittently, does this constitute a serious hazard? Explain your answer.

7
Wiring Methods

7.1   *Introduction*

Rigid metallic and nonmetallic conduit, thin-wall conduit, flexible conduit, nonmetallic sheathed cable, and armored cable are the most often employed residential and farm wiring methods recognized by the NEC. Before any given method is used, however, check your local code; it may limit the NEC specifications.

7.2   *Rigid Metallic Conduit*

Rigid metallic conduit looks like the plumber's pipe used to carry steam, gas, or water, but it is made from specially annealed steel for easy bending. Its inside surface is also very smooth in order to facilitate the easy pull-through of wires without damage to their insulation.

Available in 10-foot lengths, rigid conduit may have either a black enamel or galvanized finish. Black-enamel conduits are limited to indoor usage, but the galvanized conduits may be used indoors, outdoors, above ground or underground. When they are used in cinder fill exposed to permanent moisture, however, they must be encased in 2″ thick cement. Alternatively, they may be buried 18″ under the fill (NEC, Sec. 346-3).

Electrical conduits range from 1/2″ to 6″ in trade sizes; however, the actual physical dimensions differ appreciably from those trade sizes. See Table 7-1.

Trade size, inches	Internal diameter, inches	Internal area, square inches	External diameter, inches
0.50	0.622	0.30	0.840
0.75	0.824	0.53	1.050
1.00	1.049	0.86	1.315
1.25	1.380	1.50	1.660
1.50	1.610	2.04	1.900
2.00	2.067	3.36	2.375
2.50	2.469	4.79	2.875
3.00	3.068	7.38	3.500
3.50	3.548	9.90	4.000
4.00	4.026	12.72	4.500
4.50	4.506	15.94	5.000
5.00	5.047	20.00	5.630
6.00	6.065	28.89	6.620

Table 7-1   Dimensions of rigid metallic conduit

Cutting

Use an ordinary pipe cutter or a hack saw with 18 teeth per inch. To prevent damage to insulation, use a suitable reamer to remove all burrs on the cut edge afterwards.

Bending and threading

Conduit benders are standard equipment for an electrician. Generally, the manufacturer provides complete instructions for the use of a particular bender. Read and follow them carefully.

Table 7-2 shows the minimum permissible radius of bends in conduit conductors either with or without lead sheathing. (See Sec. 346-10). A run of conduit between outlet and outlet, between fitting and fitting, or between outlet and fitting shall not contain more than the equivalent of 4 quarter bends (360 degrees), including those bends located immediately at the outlet or fitting.

To thread rigid conduit, use a die having a taper of three-fourths of an inch per foot. Connections to boxes are made with locknuts and bushings after the conduit is threaded.

Size of conduit, inches	Minimum radius of bend in inches	
	Conductors without lead sheath	Conductors with lead sheath
0.50	4	6
0.75	5	8
1.00	6	11
1.25	8	14
1.50	10	16
2.00	12	21
2.50	15	25
3.00	18	31
3.50	21	36
4.00	24	40
4.50	27	45
5.00	30	50
6.00	36	61

Table 7-2   Minimum permissible radius of bends in conduit

Permissible number of wires

NEC Table 1 of Chapter 9, reproduced in this text as Table 7-3, gives the percent of cross-section of conduit and tubing that can be occupied by conductors. Notice that a separate listing is given for lead-covered conductors and that the table does not apply to fixture wires.

NEC Tables 3A through 3C (Chapter 9) are derived from the same Table 1 and specify the maximum number of specific types and sizes of conductors that can be enclosed in the 1/2″ to 6″ electrical trade sizes of conduit or tubing.

Number of Conductors	1	2	3	4	Over 4
All conductor types except lead-covered (new or re-wiring)	53	31	40	40	40
Lead-covered conductors	55	30	40	38	35

Table 7-3   Percent of cross section of conduit and tubing for conductors (see Table 2, Ch. 9, NEC for fixture wires)

When more than three wires are placed in conduit, their ampacity must be derated, as explained earlier in Sec. 4.7.

---

* Reprinted with permission from the 1975 National Electrical Code, Copyright 1975, National Fire Protection Association, Boston, Ma.

Support

Conduit must be firmly supported within 3 feet of each outlet box, junction box, cabinet or fitting. Either single-hole or two-hole straps can be used to support the conduit. On straight runs, the distance between supports is determined by the electrical trade size of the conduit. These requirements are computed in Table 7-4 (NEC Table 346-12).

Conduit size, inches	Maximum distance between rigid-metal conduit supports, feet
1/2 - 3/4	10
1	12
1-1/4 - 1-1/2	14
2 - 2-1/2	16
3 and larger	20

Table 7-4   Supports for rigid metallic conduit

7.3   *Rigid Nonmetallic Conduit*   PVC

Rigid nonmetallic conduit also comes in 10-foot lengths and one coupling is furnished with each length. Materials which have the physical characteristics suitable for conduit of this type are properly formed and treated fiber, asbestos cement, soapstone, rigid polyvinyl chloride (aboveground use) and high-density polyethylene (underground use). Electrical trade sizes again run from $1/2''$ to $6''$ nominally.

Use permitted (Sec. 347-2)

Rigid nonmetallic conduit and fittings approved for the purpose may be used in the following conditions and where the potential is 600 volts or less.

- In walls, floors, and ceilings

- In locations subject to severe corrosive influences (Sec. 300-6) and where subject to chemicals for which the materials are specifically approved

- In cinder fill

---

*   Reprinted with permission from the 1975 National Electrical Code, Copyright 1975, National Fire Protection Association, Boston, Ma.

- Where walls are washed frequently, the entire conduit system must be watertight. All bolts, straps, screws, etc., must be corrosion resistant or protected with approved corrosion-resistant materials.

- In dry and damp locations not prohibited by Sec. 347-3.

- For exposed work not subject to physical damage (if approved for purpose)

Use prohibited (Sec. 347-3)

Rigid nonmetallic conduit shall not be used:

- In hazardous locations except as covered in Secs 514-8 and 515-5.

- For the support of fixtures or other equipment.

- Where subject to physical damage unless approved for the purpose.

- Where subject to ambient temperatures exceeding those for which the conduit has been approved.

- For conductors whose insulation temperature limitation would exceed those for which the conduit is approved.

Miscellany

In general, the bending requirements for rigid nonmetallic conduit are the same as those given for the metallic variety.

To compensate for thermal expansion and contraction, expansion joints are sometimes required.

The number of conductors permitted in a single conduit follows the same guidelines given earlier for rigid metallic conduit.

In addition to a support within 3 feet of each box, cabinet, or other conduit termination, all rigid nonmetallic conduit must be supported as required in Table 7-5 (NEC Table 347-8).

Conduit size, inches	Maximum spacing between supports in feet	
	Conductors rated 60°C and below	Conductors rated more than 60°C
1/2 - 3/4	4	2
1 - 2	5	2-1/2
2-1/2 - 3	6	3
3-1/2 - 5	7	3-1/2
6	8	4

Table 7-5   Support for rigid nonmetallic conduit

7.4   *Thin-Wall Conduit*

Thin-wall conduit is the name commonly used in the trade to designate electrical metallic tubing (EMT). It is similar to rigid metallic conduit except that its walls are thinner, which makes it lighter and easier to handle. The largest size is 4″ tubing. It is cut and reamed by using the same methods as for rigid conduit, but the walls are not usually threaded except by special fittings approved for the purpose. Threadless connectors which consist of a body and a split ring provide tremendous pressure and electrical continuity when the nut is securely tightened. Adapters permit the use of thin-wall conduit in fittings made for rigid conduit.

All information given for rigid metallic conduit concerning the number of conductors, internal sizes, lengths, bends, and supports is applicable to EMT.

Article 348 indicates that EMT may be used for both exposed and concealed work. The use of EMT is prohibited, however, in the following cases:

- Where during installation or afterward, it may be subject to severe physical damage.

- Where protected from corrosion by enamel only.

- In cinder concrete or fill where the tubing would be subject to permanent moisture unless protected on all sides by a layer of noncinder concrete at least 2″ thick or unless the tubing is at least 18″ under the fill. Where practicable, the use of dissimilar metals throughout the system shall be avoided to eliminate the possibility of galvanic action.

- Ferrous or nonferrous EMT, elbows, couplings and fittings must not be installed in concrete or in direct contact with the earth, or in areas subject to severe corrosive influences unless they are made of a material judged suitable for the condition, or unless corrosion protection approved for the purpose is provided.

7.5   *Flexible Metal Conduit*

Often referred to as "greenfield," flexible metal conduit (Article 350) is essentially the same as the outer covering of armored cable.

Installations of flexible conduit must comply with the appropriate (or applicable) provisions of NEC Articles 300 (Wiring Methods), 334 (Metal-Clad Cable), and 346 (Nonmetallic-Sheathed Cable).

Flexible metal conduit must never be used in wet locations, unless the conductors are of the lead-covered type or of some other type specially made for the conditions prevailing in battery rooms or where rubber-covered conductors are exposed to oil, gasoline, or other materials that have a deteriorating effect on rubber. Use of flexible metal conduit in hoistways and in any hazardous location is generally, but not always, prohibited. See Sec. 350-2 for references to determine limitations.

A 3/8″ nominal trade size of flexible metal conduit may be used in lengths not in excess of 6 feet as a part of an approved assembly or for lighting fixtures. Code Table 350-3 lists the maximum number of insulated conductors acceptable in 3/8″ flexible metal conduit.

Support is required at intervals not exceeding 4′6″ and within 12″ on each side of every outlet box or fitting. For exceptions to this standard, see Sec. 350-4.

Where both the conduit and the fittings are approved for the purpose, flexible metal conduit may be used as a grounding means. *Exception:* Flexible metal conduit may be used for grounding if the length is 6 feet or less, if it is terminated in fittings approved for the purpose, and if the circuit conductors contained therein are protected by overcurrent devices rated at 20 A or less.

In most sections of the country, flexible metal conduit is seldom used. It is very useful, however, in places where it is either difficult or awkward to bend rigid conduit or where a reasonable degree of permanent flexibility is needed.

7.6   *Pull Boxes*

As you know, the NEC limits the number of bends between outlets in a run of conduit to four quarter-bends—that is, four 90° bends. Thus, the total degree of bend—including that of bends located immediately at the outlet—must not exceed 360°. If possible, of course, fewer bends should be employed since bends only complicate the installation of wires in the raceway system.

Pull boxes (Sec. 370-18) may be installed in raceway systems to reduce the number of bends in conduit. A pull box is simply a sheet metal enclosure and may be either straight or right-angled, as shown in Figs. 7-1(a) and (b), respectively. Straight pull boxes make it easier to pull wires through conduit by providing places where the electrician can reach in and help the wires along.

For raceways 3/4″ trade size or larger and containing conductors of AWG size #4 or larger, the minimum dimensions of a pull box (or junction box) installed in a raceway or a cable must comply with the following:

1. In straight pulls, the length of the box must be at least 8 times the trade diameter of the largest raceway or cable.

2. Where angle or U pulls are made, the distance from the point of raceway or cable entry to the opposite wall must be at least 6 times the trade diameter of the largest entering raceway or cable. When more than one raceway or cable enters the same side of the box, the distance between the point of entry and the opposite wall must be increased by

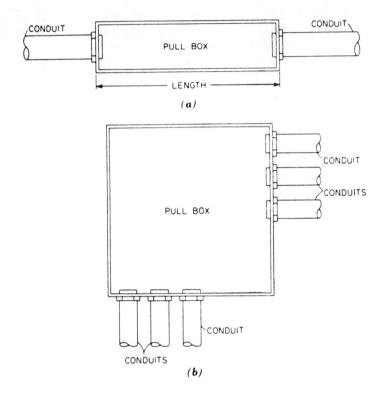

Figure 7-1   Typical straight (a) and right angle (b) pull boxes.

the sum of the trade diameters of all additional entries. The center-to-center distance between raceway entries enclosing the same conductor must not be less than 6 times the diameter of the larger raceway.

The rules for angle or U pulls are illustrated by the example shown in Fig. 7-2. Here, the pull box has two $4''$ and four $2''$ (trade diameter) entering raceways. According to the rules given above, the length of side D, and its opposite side, B, must be no less than

$$6 \times 4 + 2 + 2 = 28 \text{ inches}$$

Inspection of Fig. 7-2 shows that the same dimensions must also hold true for sides C and A.

Assuming the conductors make a $90°$ bend inside the box, and the same raceways (positionwise) enclose the same conductors, the center to center distance between raceways 3 and 4, which we will call distance X, is given by

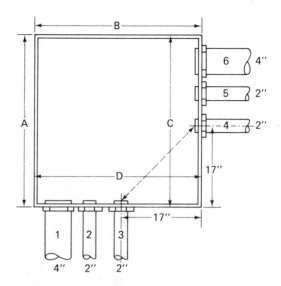

Figure 7-2   Right angle pull box dimensions. A and B represent the length of sides, C, the diagonal distance between conduits, and D, distance of conduit from corner.

$$\text{Distance X} = \sqrt{17^2 + 17^2}$$
$$= \sqrt{289 + 289}$$
$$= \sqrt{578}$$
$$= 24.04 \text{ inches}$$

Thus, the box shown in Fig. 7-2 is entirely adequate.

When the wiring is completed, pull boxes are closed by suitable covers. The covers and internal wiring must be accessible after the building is finished.

7.7   *Pulling Wire*

When raceways are completed, a continuous path exists from panelboards to pull boxes, or to junction boxes, or to outlet boxes or from outlet to outlet. You can pull the wires into the raceway system with either a snake, a fish wire, or a fish tape. These resilient pulling devices are usually made from tempered steel or plastic rope. Steel fish tapes are often wound on reels and are used as shown in Fig. 7-3. Any wires to be pulled through the conduit can be attached to the hook that is bent onto one end of the steel fish tape.

The best method to use depends, of course, on the overall length of the raceway system, the conduit and conductor sizes, the number of bends, and so

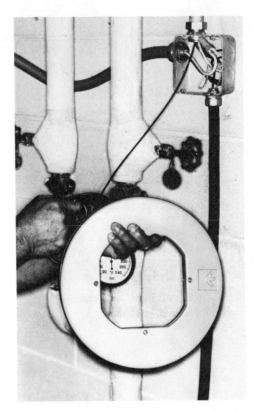

Figure 7-3   Use of steel fish tape wound on a reel (Courtesy of Ideal Industries, Inc.)

forth. Although we are concerned primarily with residential wiring, the methods described for pulling wire are, in some cases, reserved for very large jobs.

The hooked end of the fish tape is pushed into the raceway until it protrudes at an outlet. Wires which have been cut to the correct length are then attached to the fish tape so that they are pulled into the raceway as the fish tape is withdrawn. The wires should be somewhat longer than the raceway, because splices are not allowed inside the raceways. Any excess wire can always be cut off when the installation is completed. The attachment between the wires and fish tape should not be bulky, because a large attachment will make the pulling more difficult. One method of attaching the wires to the fish tape uses a socket-like device called a basket. The wire ends are inserted into this basket which is so constructed that it tightens the ends when tension is applied.

Another wire pulling method employs a compressed air gun which blows a nylon fish line through either rigid or thin-wall conduit. A plastic piston is attached to the line to receive the wire ends.

Wire pulling usually requires two men; one to pull and the other to feed the wires into the raceway and make sure there are no snarls. Powdered soapstone is often used as a lubricant to make the pulling easier.

Small wires may be attached to the fish tape and pulled by hand, but when the wires are larger, a rope or cable is first attached to the fish tape and pulled through the raceway. Then the wires are fastened to the rope or cable, which in turn is withdrawn to pull the wires into place. A hand pulling machine, which is simply a small hand-operated winch, is often helpful. Turning the handle on the winch winds the cable or rope on a drum. Gears usually lessen the pulling effort by a factor of about 50. Power pulling machines are also available for very heavy loads.

7.8 *Nonmetallic-Sheathed Cable*

As you know, there are two kinds of nonmetallic-sheathed cable; type NM has the conductors enclosed in an outer fabric braid or plastic sheath; type NMC has the conductors imbedded in solid plastic.

Code Sec. 336-3 lists the permissible and non-permissible uses of both types of nonmetallic sheathed cable.

Type NM cable may be used for both exposed and concealed work in normally dry locations. It must never be installed where it will be exposed to corrosive fumes or vapors. Nor should it be embedded in masonry, concrete, fill or plaster, or in a shallow chase in masonry or concrete and covered with plaster or similar material.

Type NMC cable may be installed for both exposed and nonexposed work in dry, moist, damp or corrosive locations, and in outside and inside walls of masonry block or tile.

Non-permissible uses for either type NM or NMC include service-entrance cable, commercial garages, theaters and assembly halls, motion-picture studios, storage-battery rooms, hoistways, any hazardous location, or when embedded in poured cement, concrete or aggregate.

Before nonmetallic-sheathed cable is installed in an outlet box, about 8 inches of the outer covering must be removed by using either a jackknife or a special cable stripper. The stripper saves time, reduces the possibility of damage to insulation, and serves as a wire gauge.

Nonmetallic-sheathed cable is anchored to switch and outlet boxes by the use of special connectors. Many boxes have built-in clamps which serve the same purpose as these connectors.

Common methods of installing nonmetallic-sheathed cable are shown in the following figure. In Fig. 7-4(a) the cable is stapled along the side of the floor joist and then passes through holes near the center of the joists. Wherever possible, holes through the joists (or any lumber) should be located at least 2″ from the edge to avoid mechanical weakening of the joist.

In Fig. 7-4(b) a running board is nailed across the joists. The cable is stapled along the bottom edge of the joist and to the running board.

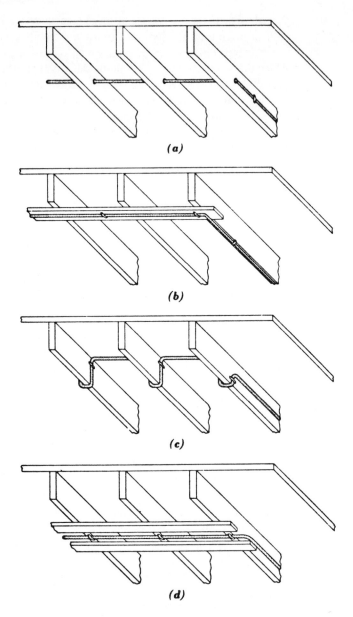

(a)

(b)

(c)

(d)

Figure 7-4   Methods of protecting cable: cable passes through holes (a), cable on running board (b), cable following building structures (c), and cable between guard strips (d)

In Fig. 7-4(c) the cable follows the building structure. This third method uses more materials and cable than necessary and the voltage drop increases. The first two methods shown in Fig. 7-4 are preferred to that shown third.

When attics are wired, guard strips like those shown in Fig. 7-4(d) are used to protect cables that cross rafters and joists. The guard strips must be at least as high as the cable is thick. Where the cable is more than 7 feet above the attic floor, guard strips are not needed. Protection is needed only for 6 feet around the attic access opening if there is no permanent stairway or ladder.

When installing cables, always follow the approximate contour of the building. Do not take short cuts across open spaces. When nonmetallic-sheathed cable is run at an angle with the joists in unfinished basements, cables no smaller than two #6 or three #8 may be fastened directly to the lower edges of the joists. For smaller cables use the method shown in Figs. 7-4(a) and (b). Cables of all sizes run parallel to joists and must be fastened to the sides or edges of the joists.

Nonmetallic-sheathed cable must be anchored every $4'6''$ and in any case within $12''$ of an outlet box. Avoid the use of staples since this may cause damage to the insulation. No bends should have a radius of less than 5 times the diameter of the cable.

7.9 Armored Cable

Armored (metal-clad) cables are available in type MC or AC series, with acceptable metal covering. Since type MC contains conductors of AWG #4 and larger (power cables), our discussion will be limited to type AC use in branch circuits and feeders.

Armored cable of the AC type may be used in dry locations. Where walls are exposed to excessive moisture or dampness or are below grade line, type ACL (lead-covered cable) must be used. (See Article 334 for complete information.)

A two-conductor armored cable is shown in Fig. 7-5. Note that a bonding strip is in intimate contact with the armor. As explained earlier, the purpose of this bonding strip is to reduce the resistance of the wire. An insulating bushing is slipped inside the armor to protect the insulation from

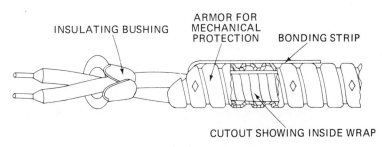

Figure 7-5   Two-conductor armored cable

sharp edges on the armor. The cutout shows the inside paper wrap. When the insulating bushing is being installed, keep the bonding strip outside the bushing.

Armored cable may be cut either with a hacksaw or an armor cutter. Any hacksaw used for this purpose should have 32 teeth per inch. Hold the hacksaw at right angles to the armor—not at right angles to the cable. Take care not to damage the insulation of conductors inside the cable. After you cut the armor, twist it to separate the parts. Once the armor is separated, it can be pulled off to uncover the bonding strip and paper wrapping.

EXERCISES

7-1/   Describe the appearance and finish of rigid metallic conduit.

7-2/   In what way is the use of black-enamel rigid metallic conduit limited?

7-3/   In what lengths and in what trade sizes are rigid metallic conduit available?

7-4/   How is rigid metallic conduit cut?

7-5/   What restriction is placed on the bending of rigid metallic conduit?

7-6/   When installed in rigid metallic conduit, must the ampacity of conductors be given consideration? Explain in detail.

7-7/   What are the support requirements for rigid metallic conduit?

7-8/   Where may rigid nonmetallic conduit be used?

7-9/   Where may rigid nonmetallic conduit not be used?

7-10/   Where thermal expansion and/or contraction are experienced, what precaution must be taken with rigid nonmetallic conduit?

7-11/   In what electrical trade sizes is thin-wall conduit available?

7-12/   Is EMT threaded in the same manner as rigid metallic conduit? Explain fully.

7-13/   Where is the use of EMT prohibited?

7-14/   In what types of installations is "greenfield" handy?

7-15/   What restrictions are placed on the use of flexible metal conduit?

7-16/   How must flexible metal conduit be supported?

7-17/   What is the purpose of a pull box?

7-18/   If a straight pull box is used with, say 3″ conduit, what is its required length?

7-19/   A right-angle pull box has one 2″ conduit and three 1″ conduits attached on each of two sides. What must be the minimum size of the box? What is the distance between the centers of the two nearest 1″ conduits? Assuming that the pull box is square and the two conduits are the same distance from the corner, determine distance D, Fig. 7-2.

7-20/   In a small residential installation employing rigid metallic tubing, describe how you would normally pull the conductors into the raceway system.

7-21/   What are the two types of nonmetallic-sheathed cable, and how may they be used?

7-22/   Describe three techniques for installing nonmetallic-sheathed cable.

7-23/   What precautions must be taken when nonmetallic-sheathed cable is used to wire an attic that is less than 7 feet high?

7-24/   What is the purpose of the bonding strip which is in intimate contact with the outside sheath of armored cable?

7-25/   Describe how you would cut armored cable with a hacksaw.

8

Sizing the Service Entrance and Branch Circuits

8.1 *Introduction*

Since the service, service equipment, and systems ground provide protection for the rest of the wiring system, most inspections start at this point.

Once the inspection of the service, service equipment and systems ground is completed, the inspector will turn his attention to branch circuits and to feeders—where they exist.

By definition (NEC, Article 100), a feeder is the circuit conductors between the service equipment and the branch circuit overcurrent devices. Since, in a residential wiring system, both the main and branch circuit overcurrent devices are located in the same cabinet, there are no feeders.

Again, by NEC definition, a branch circuit is that portion of the wiring system between the final overcurrent device protecting the circuit and the outlet or outlets. In determining where a branch circuit actually begins, notice that this definition specifically states ". . .between the final overcurrent device protecting the circuit . . . ." Thus, a device not approved for branch-circuit protection, such as a thermal cutout or motor-overload protective device, is not considered to be the starting point of a branch circuit.

In order to plan a wiring system, it is necessary to make certain calculations, such as those demonstrated in this chapter. First, however, let us take a look at the symbols used in wiring diagrams and the types of building plans you will encounter.

---

* Reprinted with permission from the 1975 National Electrical Code, Copyright 1975, National Fire Protection Association, Boston, Ma.

8.2   *Symbols*

Electrical symbols that commonly appear on the architectural drawings of residences are shown in Fig. 8-1. Study these symbols carefully and refer to them as needed. Notice in particular that round and square symbols designate outlets served by full voltage and low voltage, respectively.

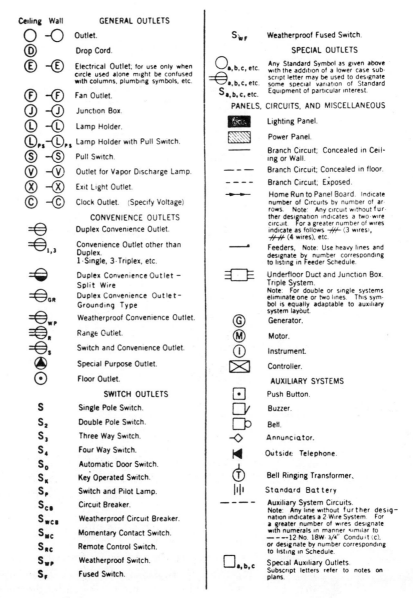

Figure 8-1   Common symbols

8.3   *Getting Familiar with Building Plans*

The use of symbols is illustrated in the very simple wiring plan for a two-car garage shown in Fig. 8-2. Duplex convenience outlets, lamp holders, and switches are indicated by their respective symbols. The dashed lines connecting the devices indicate that exposed wiring is used for the connections between the devices. The lighting panel is shown without any indication of whether it contains switches and fuses or circuit breakers; therefore, wiring plans should include notes giving additional information such as wire sizes, number of circuits, and type of equipment.

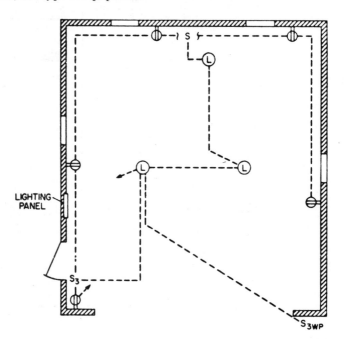

Figure 8-2   Garage wiring plan

You can study the wiring layout for a small house with a basement in Fig. 8-3. The laundry facilities and lighting panel are assumed to be in the basement; and for simplicity, the basement plan with its wiring layout is not shown.

In the kitchen, receptacles are provided for a refrigerator, clock, iron, toaster, and other small appliances. A special outlet, marked S_p, is provided for ironing; it has a switch and a pilot lamp. Special outlets are provided for the range (R), range hood (RH), garbage disposal (GD), and dishwasher (DW). Note that the two small-appliance circuits are shared by the kitchen and dining room and that they do not enter other rooms.

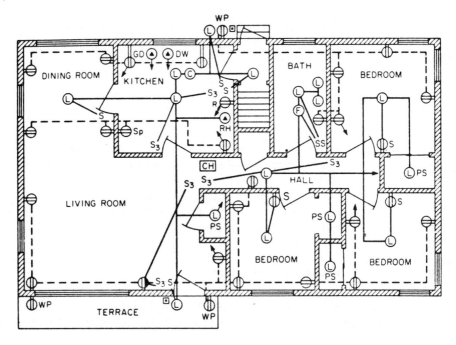

Figure 8-3   Wiring layout for a small house

Adequate receptacles are installed in the other rooms. They are spaced at approximately equal distances (always less than 12 feet apart). Each bedroom has a combination switch/receptacle outlet and a ceiling light for general illumination. Each closet has a light controlled by a pull-chain switch. The receptacle at the entrance door is convenient for connecting a vacuum cleaner or other small appliance.

The bath has a ceiling light for general illumination and special lights at the mirror. It also has a separate exhaust fan.

Near the front door there is a split-wired duplex receptacle. The top half of the receptacle is wired to two 3-way switches, and the bottom half is always alive. The terrace has two weatherproof receptacles for portable lamps, decorative lighting, etc.

Three-way switches are used in the hall, kitchen and living room to reduce the amount of walking required for control. Notice the push-button switches at both doors. These switches control the hall chimes.

The lines connecting the outlets in Fig. 8-3 represent runs of cable. The arrowheads on the circuits indicate the *home runs*, which are the cables leading to the service entrance. The number of circuits can be determined by counting the arrowheads. The solid lines indicate circuits concealed in the ceilings or walls and the broken lines indicate exposed wiring in the basement.

At first, reading and understanding wiring diagrams may be somewhat difficult, but with practice, you will find that this part of the electrician's job is not difficult. A good practice exercise is to make two wiring diagrams of your own home; one as it actually exists, and one as you would like to have it.

8.4   *Determining the Load—General*

When the wiring for a house is planned, the requirements of the NEC and any local code must be used as a guide. Keep in mind that these requirements represent a *minimum*. Good wiring practice may require additional or higher-capacity circuits. The computed load should represent all the power required for lighting, small appliances, and other loads such as water pumps, water heaters, space heaters, clothes driers, and air conditioners. It is also very important to provide for future additional appliances which will increase the load on the electrical system.

8.5   *Lighting Load*

NEC Table 220-2(b) specifies that the lighting load be calculated on the basis of 3 watts per square foot (3 W/ft^2) of occupied space. The floor area is computed from the outside dimensions of the building, apartment, or area involved, and by the number of floors. Open porches, attached garages, and unfinished or unused spaces (unless adaptable for future use) are not included in this computation.

The 3 W/ft^2 unit value noted above is based on *minimum* load conditions and 100 percent power factor, and may not provide sufficient capacity for the installation contemplated.

In view of the trend toward higher-intensity lighting systems and increased loads due to more general use of fixed and portable appliances, each installation should be calculated for the load *likely* to be imposed along with a capacity-increased factor to ensure safe operation.

Where electric-discharge lighting systems are to be installed, a high-power-factor-type light should be used to avoid the need for conductors having greater ampacity.

The *continuous* load—that is, a load where the maximum current is expected to continue for three or more hours—supplied by a branch circuit must not exceed 80 percent of the branch-circuit rating. *Exceptions*: (1) Where the assembly, including the overcurrent device protecting the branch circuit is approved for operation at 100 percent of its rating, the continuous load supplied by the branch circuit may equal the ampacity of the branch-circuit conductors. (2) Where branch circuits are derated, branch-circuit loads must not exceed the derated ampacity of the conductors. (Review Sec. 4.7 of this text.)

To determine the installation requirements of a typical residential load, assume a one-story house with a full basement. The outside dimensions of the ranch-type house are 30 feet by 60 feet and it is considered likely that the basement will someday be finished. Then, the total area on which to base your calculations is $30 \times 60 \times 2 = 3600$ ft^2. Allowing 3 W/ft^2, the lighting load would be $3 \times 3600 = 10,800$ W. To determine amperage, now transpose the power equation to $I = P/E$. Using the load requirement determined above, and assuming a supply voltage of 115 V,

$$I = \frac{P}{E} = \frac{10,800}{115} = 93.9 \text{ A}$$

For practical purposes, the amperage requirement is taken as 94 A.

Each circuit is usually designed to carry either 15 or 20 A. If the local code prohibits the use of 15 A circuits, then you would use 20 A circuits. The number of circuits needed is determined by dividing the total current by the required amperage and rounding the answer off to the next higher whole number. Assuming 15 A circuits and a total current requirement of 94 A, the "sample" house would need

$$\frac{94}{15} = 6.3 \text{ circuits}$$

Rounded off to the next higher whole number, 6.3 becomes 7. Thus, you can see that seven 15 A circuits would be needed. Alternatively, the number of 20 A circuits required is

$$\frac{94}{20} = 4.7 \text{ circuits}$$

Rounded off to the next higher whole number, 5 circuits would be needed.

With 15 A circuits you should use AWG #14 copper wire; and with 20 A circuits you must use AWG #12 copper wire.

The load should always be distributed over several circuits to prevent the whole house from being darkened by the blowing of one fuse or the tripping of one circuit breaker. Also, to the greatest extent practical, the load should be divided evenly between branch circuits.

Another NEC recommended way for you to determine the minimum number of lighting circuits needed is to allow one circuit for each 500 ft^2 of floor area, however, actually allowing one circuit for each 400 ft^2 of floor area provides a good safety margin. Be sure to install fixed lights and duplex receptacles meant for floor lamps, TV sets, and similar loads on separate circuits.

8.6   *Small-Appliance and Laundry Loads*

The small appliances considered in this group do not include any fixed appliances, such as garbage disposals, built-in dishwashers, and garbage compactors.

For the small-appliance load in kitchen, pantry, family room, dining room, and breakfast room of dwelling occupancies, two or more 20 A branch circuits must be provided for all receptacle outlets, in addition to those required for lighting. Such small-appliance branch circuits must have no other outlets.

Receptacle outlets supplied by at least two appliance-receptacle branch circuits must be installed in the kitchen.

Each small-appliance branch circuit must be able to supply a minimum of 1500 W.

A laundry load of 1500 W is assumed for a typical residence. The load is supplied by a separate 20 A branch circuit which terminates with one duplex receptacle located within six feet of the intended location of the washing machine. If an electric clothes dryer is to be used, another circuit is installed in the laundry area.

8.7   *Electric Ranges*

Electric ranges are supplied by a separate 240 V circuit. Before we say more about ranges, however, let us define one very important term—*demand factor.*

Demand factor is the ratio of the maximum demand of the system, or any part of the system, to the total connected load of the system or of that part of the system under consideration.

Because all loads on a system are rarely turned on simultaneously, the maximum demand is considered to lie somewhere between the maximum connected load and the actual usage. The demand factor is expressed as a percentage of the maximum load and is used in practice to determine the size of conductors and/or overcurrent devices needed for a particular application. The demand factor is determined by a series of tests, and the resulting information appears in Table 8-1 (NEC Table 220-19).

| | Maximum Demand | Demand Factors | |
Number of Appliances	Column A (not over 12 kW rating)	Column B (less than 3½ kW rating)	Comumn C (3½ kw to 8¾ kW rating)
1	8 kW	80%	80%
2	11 kW	75%	65%
3	14 kW	70%	55%
4	17 kW	66%	50%
5	20 kW	62%	45%
6	21 kW	59%	43%
7	22 kW	56%	40%
8	23 kW	53%	36%
9	24 kW	51%	35%
10	25 kW	49%	34%

Number of Appliances	Maximum Demand	Demand Factors	
	Column A (not over 12 kW rating)	Column B (less than 3½ kW rating)	Column C (3½ kw to 8¾ kW rating)
11	26 kW	47%	32%
12	27 kW	45%	32%
13	28 kW	43%	32%
14	29 kW	41%	32%
15	30 kW	40%	32%
16	31 kW	39%	28%
17	32 kW	38%	28%
18	33 kW	37%	28%
19	34 kW	36%	28%
20	35 kW	35%	28%
21	36 kW	34%	26%
22	37 kW	33%	26%
23	38 kW	32%	26%
24	39 kW	31%	26%
25	40 kW	30%	26%
26-30	15 kW plus 1 kW for each range	30%	24%
31-40		30%	22%
41-50	25 kW plus 3/4 kW for each range	30%	20%
51-60		30%	18%
61 & over		30%	16%

Table 8-1   Demand loads for household electric ranges, wall-mounted ovens, counter-mounted cooking units and other household cooking appliances over 1-3/4 kW rating. Column A is to be used in all cases except as otherwise permitted in Note 3 below.

Note 1. Over 12 kW through 27 kW ranges all of same rating. For ranges, individually rated more than 12 kW but not more than 27 kW, the maximum demand in Column A shall be increased 5 percent for each additional kW of rating or major fraction thereof by which the rating of individual ranges exceeds 12 kW.

Note 2. Over 12 kW through 27 kW ranges of *unequal ratings.* For ranges individually rated more than 12 kW and of different ratings but none exceeding 27 kW an average value of rating shall be computed by adding together the ratings of all ranges to obtain the total connected load (using 12 kW for any range rated less than 12 kW) and dividing by the total number of ranges; and then the maximum demand in Column A shall be increased 5 percent for each kW or major fraction thereof by which this average value exceeds 12 kW.

Note 3. Over 1-3/4 kW through 8-3/4 kW. In lieu of the method provided in Column A, it shall be permissible to add the nameplate ratings of all ranges rated more than 1-3/4 kW but not more than 8-3/4 kW and multiply the sum by the demand factors specified in Column B or C for the given number of appliances.

Note 4. Branch-Circuit Load. It shall be permissible to compute the branch-circuit load for one range in accordance with Table 220-19. The branch-circuit load for one wall-mounted oven or one counter-mounted cooking unit shall be the nameplate rating of the appliance. The branch-circuit load for a counter-mounted cooking unit and not more than two wall-mounted ovens, all supplied from a single branch circuit and located in the same room, shall be computed by adding the nameplate rating of the individual appliances and treating this total as equivalent to one range.

Note 5. This table also applies to household cooking appliances rated over 1-3/4 kW and used in instructional programs.

8.8   *Motor Loads*

Loads such as air conditioners are considered as motor loads. The nameplate current rating of the electric motor is changed to an equivalent number of kilowatts using the formula $P = EI$. An air conditioner which has a nameplate rating of 7 A at 230 V, for example, requires

$$7 \times 230 = 1610 \text{ W} = 1.61 \text{ kW}$$

Sometimes it is convenient for your calculations to use the current drawn by the motor. If a branch circuit supplies only one motor, then branch-circuit ampacity should be at least 1.25 greater than the motor current. When the branch circuit supplies several motors, multiply the largest current by 1.25 and add the result to the sum of the other motor currents. Consider the case when a branch circuit supplies three motors drawing respectively 6, 9, and 12 A. The 12 A current is multiplied by 1.25 before adding it to the other currents:

$$12 \times 1.25 = 15 \text{ A for the 12 A motor}$$

$$6 + 9 + 15 = 30 \text{ A total current}$$

Therefore, a 30 A branch circuit is needed to supply these three motors. The same method may be used to calculate current and determine the size of service-entrance equipment.

Other common motor loads which should be considered are garbage disposals, water pumps, heating system motors, and permanent attic fans.

8.9   *Other Loads*

The power used by appliances, such as clothes dryers, dishwashers, and electric space heaters, is calculated according to the nameplate ratings of the appliances. Only the more common appliances have been mentioned in load calculations. Other loads which may have to be considered include those used for hobbies and/or in a small business.

8.10   *Alternate Loads*

In calculating the size of service-entrance wires, you will, of course, consider the fact that certain load combinations are not likely to occur simultaneously. Air conditioners, for example, are not likely to be operating at the same time as space heaters. Use the larger of the two or more loads then to calculate load requirements. If the air conditioner uses, say, 2 kW and the electric space heater uses 10 kW, you would only consider the space-heater load. If a space heater supplements a heat pump, however, remember that the space heater *plus* the heat pump makes up the total load since they would be used simultaneously.

8.11   *Sizing the Service Entrance*

Once the total load in kW is determined, the service entrance size is calculated. First, from the total number of kW, compute the total current requirement. This requirement determines the smallest permissible sizes for the service-entrance wires and for the *panelboard*—the metal enclosure which houses the fuse-switch arrangements or circuit breakers. Because all loads are not likely to be used at the same time, a demand factor like that used for electric ranges can be applied to the total load. One approved method is to add up all the loads, excluding the electric range. For the first 3 kW use a 100 percent demand factor and for the remainder, a 35 percent demand factor. To this total, add the range load computed in the manner explained in Sec. 8.7. A second approved method is to add all loads, including the electric range's nameplate rating, then use a demand factor of 100 percent for the first 10 kW and a 40 percent demand factor for the remainder.

The computation of service-entrance size is best illustrated by the examples presented below. NEC's Chapter 9 includes similar example calculations. You should study them and become thoroughly familiar with the methods used for load determination.

8.12   *Example Problems*

Problem 1

A single-family, ranch-style home has a floor area of 2,000 square feet and is built on a cement slab and an oil burner used in the heating system has two 1/4 hp motors, each drawing 6 amp. What is the minimum size requirement for the service-entrance conductors?

Solution

The lighting load, based on 3 W/ft^2 is 2000 × 3 = 6000 W = 6 kW. A small-appliance load of 3 kW and a laundry load of 1.5 kW, are assumed to meet with the minimum requirements of the NEC. Each furnace motor uses a power of (6 × 115)/1000 = 0.69 ≈ 0.7 kW. The total load is reckoned thus:

Load	Power in kW
Lighting	6.0
Small appliances	3.0
Laundry	1.5
Furnace	1.4
Total	11.9

For the first 3 kW of the load, the demand factor is 100 percent, for the remaining 8.9 kW, the demand factor is taken as 35 percent, or 0.35 × 8.9 = 3.1 kW. Then, the total load is 6.1 kW.

Since 6.1 kW = 6100 W and $I = P/E$, the current in the service-entrance conductors for a 115/230 V three-wire system is

$$\frac{6100}{230} = 26.6 \approx 27 \text{ A}$$

Although the current is only 27 A, a 100 A service entrance must be used because the allowable minimum number of circuits is five.

Problem 2

Now consider the residence described in Problem 1 a total-electric house using 12 kW of electric space heating with individual thermostats in five rooms, a 5 kW clothes dryer, a 1.5 kW dishwasher, a 2 kW water heater, a 12 kW electric range, and an air-conditioning load of 6 kW, what is the minimum size that can be used for the service-entrance conductors?

Solution

Since the air-conditioning load is less than the electric space-heating load, it is set aside, and the total is reckoned as follows:

Load	Power in kW
Lighting, at 3 W/ft^2	6.0
Small appliances	3.0
Laundry	1.5
Electric space heating	12.0
Clothes dryer	5.0
Dishwasher	1.5
Water heater	2.0
Electric range (nameplate rating)	12.0
Total	43.0

The demand factor for the first 10 kW is 100 percent and, for the remaining 33 kW, 40 percent. Since $0.40 \times 33 = 13.2$ kW, the total load is $10 + 13.2 = 23.2$ kW $= 23,200$ W, and the current in the service entrance wires is

$$\frac{23,200}{230} = 100.9 \approx 101 \text{ A}$$

For such an installation, choose service-entrance equipment and wire of the next larger size. For example, Number 2/0 copper wire with type T insulation rated at 145 A is suitable. Standard panelboard ratings are 140 A for circuit breaker types and 200 A for fusible types.

In Problem 2, the electric space heating was derated because each room had its own individual thermostat. If there are fewer than *four* separate thermostats, no derating is used. If, for example, only three separate thermostats were used, the load would be 10 kW at 100 percent demand factor, plus 12 kW for the heating at 100 percent demand factor, plus 8.4 kW (21 kW at 40 percent demand factor) for a total of 30.4 kW. The service entrance current would be

$$\frac{30,400}{230} = 131.2 \approx 131 \text{ A}$$

A 140 A service entrance could be used in both cases, but note that the non-derated heating unit increases the current by about 31 percent.

8.13   *Other Selection Factors*

The 100 A service entrance is the smallest size normally used. The service entrance must have a current capacity equal to or greater than the calculated load. A capacity that is larger than the calculated load is preferred, because an increased load may be added later. A 60 A service entrance may be used only when fewer than six two-wire branch circuits are used and the computed load is less than 10 kW.

Because most houses have occupied areas greater than 1000 square feet, a minimum of three circuits are needed for general lighting and appliances. This is based upon the NEC recommendation of at least one circuit for every 500 square feet. Remember that you must allow two circuits for small appliances and one for a laundry. Even without considering heating or air conditioning, you can see that six circuits are needed. That number exceeds the five circuits permitted with a 60 A service entrance. Consequently, the 100 A service is the smallest size in common use. The Local electrical codes must also be considered because some do not permit the use of service entrances with ratings of less than 200 A.

8.14   *Branch Circuits*

The number of branch circuits affects the size of the service entrance because the service-entrance conductors must match the service-disconnect and the service-overcurrent devices. If many branch circuits are used, a large panelboard is required. Understandably, large panelboards have high-current ratings.

The usual procedure is to provide a branch circuit for each of the listed loads:

1.   Electric range

2.   Automatic clothes washer

3.   Clothes dryer

4.   Electric water heater

5.   Dishwasher

6.   Garbage disposal

7.   Water pump

8.   Motor on oil-burning furnace

9.   Motor on blower in furnace

10.   Permanently connected motors with ratings greater than 1/8 hp

11.   Permanently connected appliances with ratings greater than 1000 W

12.   Any automatically-started motor such as a fan

The circuits can be either 115 or 230 volt. Circuits rated at 230 volts are preferred because smaller wires can be used. Also, the voltage drop in the wires is lower.

When more than one receptacle is used on a branch circuit, the circuit capacity is limited. For example, a 15 A branch circuit can not supply a portable appliance requiring more than 12 amp or 1380 W. A 20 A circuit is limited to 16 A or 1840 W. The portable appliance should not use more than 80 percent of the circuit capacity. When a branch circuit supplies both portable and permanently connected appliances, the fixed load must not exceed 50 percent of the circuit rating. Therefore, the highest fixed load for a 15 A circuit is 7.5 A or 863 W. A 20 A circuit can supply fixed appliances using up to 10 A. This limits the circuit to fixed-appliance loads up to 1150 W.

Branch circuits which are rated at 30, 40, or 50 A can not be used for lighting or for small-appliance circuits in residences. Their use is limited to fixed appliances such as electric water heaters and fixed space heaters. The NEC specifies that the maximum load on a 30 A circuit shall not exceed 80 percent of circuit ampacity. Thus a 30 A circuit can supply a load which requires up to 0.8 × 30 = 24 A. The 40 and 50 A circuits are restricted to supplying electric ranges.

How many branch circuits should you install in a house? As you can see from the detailed considerations above, there is no standard answer because the number of circuits depends upon the size of the house and the expected electrical load. As a general rule, there should be more branch circuits than are minimally required by the National Electrical Code. The presence of extra circuits permits the addition of other electrical equipment in the future. Extra circuits also make the electrical system more convenient because each circuit carries less load and that reduces the chances of blown fuses or tripped circuit breakers. Finally, the load is better balanced between circuits, so that the power waste for heating wires is reduced.

8.15   *Feeder Circuits*

The service-entrance conductors of single-family residences can be considered as feeders for certain loads. A 75 percent demand factor may be applied when four or more fixed-appliance loads are connected to these conductors. However, the appliances must be other than electric ranges, clothes dryers, air conditioners, or space heaters. If a residence contains four loads such as water heaters, fixed dishwashers, garbage disposals, furnace motors, or water pumps, the demand factor can be used. The same demand factor can be also applied to similar loads in multi-family dwellings. The following example problem will clarify the use of the demand factor.

Problem 3

What is the current in the ungrounded line when a 230 V 2500 W water heater, a 1/2 hp, 115 V garbage disposal, a 1/2 hp, 230 V water pump, a 1/4 hp, 115 V furnace motor, and a 1/4 hp, 115 V attic fan are fed by a 115/230 V feeder? The motor nameplate currents are 4.8 for the 1/4 hp motors, 6.6 A for the garbage disposal, and 3.6 A for the water pump.

Solution

The current drawn by the water heater is

$$I = \frac{P}{E} = \frac{2500}{230} = 10.9 \text{ A}$$

Because the garbage disposal draws the largest motor current, its current is increased by 1.25 to determine the line current:

$$1.25 \times 6.6 = 8.25, \text{ or } 8.3 \text{ A}$$

To balance the load, the furnace motor and the attic fan should be connected to different ungrounded conductors. It is sufficient to compute the current in the ungrounded line which has the larger load. In this example, the line feeding the garbage disposal has the larger load. The total current in this line is calculated as follows:

Load	Current, A
Water heater	10.9
Garbage disposal	8.3
Furnace motor	4.8
Attic fan	0.0
Water pump	3.6
Total	27.6

When the demand factor of 75 percent is considered, the line current is

$$0.75 \times 27.6 = 20.7 \text{ A} \quad \text{Answer}$$

NOTE. When feeder circuits are used, the line current calculated for the service-entrance conductors is reduced—from 27.6 to 20.7 amp in this example. When the size of the service entrance is determined, the other loads, such as lighting, small appliances, and electric range, are calculated as before. The feeder current is added to the total current of other loads to determine the total current in the service entrance.

8.16   *Selecting the Panelboard*

Once the total current and the number of circuits have been determined, the panelboard can be selected. Its current rating can not be less than the computed load current. Also, it must protect the service entrance conductors from excessive current. A typical panelboard in newer homes uses circuit breakers. The main circuit breaker, which has a rating equal to that of the panelboard, is located at the top of the panelboard. Below the main circuit breaker are the two-pole circuit breakers for protecting 230 V circuits, and then single-pole circuit breakers, for protecting 115 V circuits. For safety, all the circuit breakers are mounted in a dead-front cabinet. That means that there are no exposed terminals or uninsulated wires showing and the cabinet door is labeled with an index on which the use of each circuit breaker is written after installation. Sometimes two circuit breakers are connected in parallel on the watthour-meter side to provide special ratings. For example, two 70 A circuit breakers can be used to make a 140 A panelboard. Service circuit breakers are rated at 30, 40, 50, 60, 70, 100, 125, 150, 175, and 200 A. Equipment larger than 200 A is available for larger installations.

A panelboard may also have pull-out switches and plug fuses. The standard ratings are 30, 60, 100, and 200 A. The main lighting switch is a pull-out switch which holds two cartridge fuses. Depending upon the internal connections in the panelboard, the main switch may protect the other pull-outs and the plug fuses or only the plug fuses. There may also be pull-outs for the range, water heater, and clothes dryer. Then come fuse holders for plug fuses. The plug fuses protect 115 V branch circuits used for lighting, laundry, and small appliances. Remember that, in new installations, adapters for type S fuses must be inserted in the fuse holders designed for plug fuses.

Each pull-out switch holds two cartridge fuses for a particular size of fuse. For example, a fuse holder made for fuses between 35 and 60 A will not accept fuses 30 A and smaller or 70 A and larger. Each pull-out can be inserted in two positions. In the correct position, the current flows through normally; whereas in the reverse position the pull-out acts as an open switch. It is easy to determine the positions because of the markings molded into the pull-out switch.

When you select the panelboard for a house, be wise and select one larger than is actually need at the time of installation so that there will be room for spare branch circuits. No one can foresee all the equipment that a householder may wish to add in the future, but some can be anticipated in this way. Also, the panelboard should conform to local code custom. Contact your power supplier or electrical inspector to determine what equipment is acceptable.

EXERCISES

8-1/   How would you determine where a branch circuit begins?

8-2/   In building plans, how does the architect indicate (a) exposed and (b) covered wiring used for connection between devices?

8-3/   What symbols are used to designate outlets served by (a) full voltage and (b) low voltage?

8-4/ What letter symbols are used to designate (a) a range, (b) dishwasher, (c) garbage disposal, and (d) range hood?

8-5/ In building plans, how are cables leading to the service entrance indicated? How can the number of branch circuits be determined?

8-6/ For wiring a house, is it wise to use NEC and/or local code requirements? If so, why?

8-7/ If the lighting load for a house is based on 3 W/ft^2, does this ensure that the system will privide sufficient capacity for the installation contemplated? Explain your answer.

8-8/ What precaution should be taken in planning a house-wiring layout if the use of electric-discharge lighting is contemplated? Explain your answer.

8-9/ What is meant by the expression "continuous load"?

8-10/ In general, what is the continuous-load limitation on a branch circuit?

8-11/ In a two-story house having a finished basement, the outside dimensions are 32 feet by 58 feet. What is the lighting load? What is the total current requirements with a supply voltage of 115 V?

8-12/ For the house in Example 8-11, how many (a) 15 A and (b) 20 A branch circuits are required for the lighting system?

8-13/ Why should the lighting load always be distributed over several branch circuits?

8-14/ Using the alternative of allowing one lighting circuit for each 500 ft^2 of floor space, how many circuits would be required for the house described in Example 8-11?

8-15/ What are the branch-circuit requirements for the small-appliance load in the kitchen, pantry, family, dining, and breakfast rooms of dwelling occupancies?

8-16/ How much wattage must each small-appliance branch circuit be able to supply?

8-17/ What laundry load is assumed in a typical residence and how must this load be supplied?

8-18/ What provision must be made if the use of an electric clothes dryer is also contemplated?

8-19/ What is meant by the expression "demand factor"?

8-20/ What is the voltage requirement for an electric range?

8-21/ How is demand factor (a) expressed and (b) determined?

8-22/ For a range rated at 10 kW, what maximum demand is assumed?

8-23/ For a range rated at 7.5 kW, what maximum demand is assumed?

8-24/ For a total range load of 11 kW, what is the total demand?

8-25/ What load is imposed by an air conditioner carrying a nameplate rating of 6 A at 230 V?

8-26/ Suppose a branch circuit supplies one motor drawing 8 A. What is the required ampacity of the branch circuit?

8-27/ Suppose a branch circuit supplies two motors drawing respectively 8 and 10 A. What is the required ampacity of the circuit?

8-28/ In calculating the total load on a wiring system, would you include both a washer and dryer? Explain your answer.

8-29/   Describe two methods for determining total load requirements for a residential wiring system.

8-30/   A ranch-style home built on a cement slab has a floor area of 1800 ft^2; an oil burner used in its heating system hs two 1/4 hp motors. What minimum size is required for the service-entrance conductors?

8-31/   Assume that the home in Example 8-28 is "all electric" and uses 20 kW for electric heating, with individual thermostats in four rooms, a 6 kW clothes dryer, a 1.5 kW dishwasher, a 3 kW water heater, a 9 kW range, and a 7 kW air conditioner. What is the minimum size required for the service-entrance conductors?

8-32/   When may the heating load be derated for electric space heating?

8-33/   Normally, what is the smallest size allowed for a service entrance? Why?

8-34/   Under what special circumstances may a service entrance smaller than 100 A be used?

8-35/   Name 10 loads for which a separate branch circuit is normally used. With such loads, is it better to use a 115 V or a 230 V supply source? Give the reason for your choice.

8-36/   For portable appliances, what is the maximum wattage that can be supplied by (a) 15 A and (b) 20 A branch circuits supplied by a 115 V source?

8-37/   When a branch circuit supplies both portable and permanently connected appliances, how much of the circuit rating, expressed as a percentage, can apply to the fixed load?

8-38/   What is the maximum permissible load on a 30 A branch circuit?

8-39/   If the service-entrance conductors of a single-family residence are considered as feeders for certain loads and four or more fixed-appliance loads are connected to these conductors, what demand factor may be applied? What appliances must be excluded from such calculations?

8-40/   Describe typical panelboards using (a) circuit breakers and (b) pull-out switches and plug fuses.

8-41/   For residential use, what are the standard ratings for circuit breakers?

8-42/   What are the standard ratings for pull-out switch and plug-fuse type panelboards?

8-43/   What factors should be considered in selecting a panelboard?

Part
4

9

Installing the Service Entrance and Ground

9.1   *Preparation for Wiring*

Before any electrical wiring is started, the complete system is designed by either an architect, a contractor, or an electrician. If the design is completed before the electrician starts a job, it is only necessary for him to follow the construction and wiring plans in order to install the specified wires and equipment at the correct locations. For many small installations, however, the design is either nonexistent or sketchy and the electrician is called upon to design a system that will meet the requirements of whatever code prevails—either local or NEC.

Schematic diagrams

Because buildings are seldom wired alike, it is not possible to draw a general plan that will apply to all installations. Certain main features are common to all wiring plans, however, regardless of the use to which the building is put.

Wiring plans for the service entrance of a typical residence are shown in Fig. 9-1(a) in combined pictorial-schematic form and in Fig. 9-1(b) in pure schematic form. In each case a 115/230 V, single-phase, three-wire circuit is assumed. Figure 9-1(a) is self explanatory. Notice in Fig. 9-1(b) however, that a

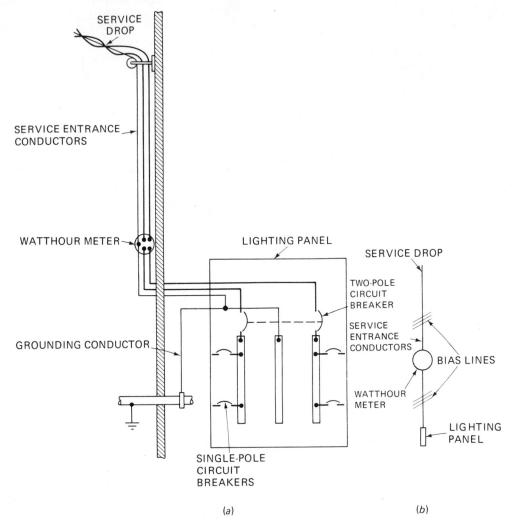

Figure 9-1   Wiring plans for the service entrance of a typical residence: (a) detailed, and (b) schematic forms

single straight line indicates both the service drop and the service-entrance conductors. The fact that there are three conductors is indicated by the bias lines. Also in Fig. 9-1(b), the kWh meter is simply indicated by a circle and no internal details of the panelboard and systems are shown. Diagrams of the type shown in Fig. 9-1(b) are often referred to as one-line schematic diagrams. Unfortunately, the recommended uniform method is not always followed in the preparation of such diagrams. In some one-line schematic diagrams, for example, the bias lines are omitted and other methods are used to specify the number of wires.

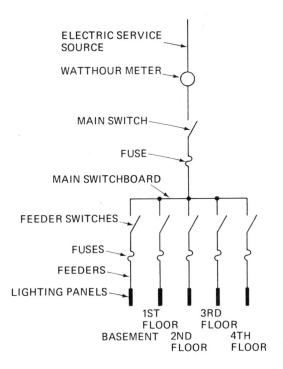

Figure 9-2   A modified form of one-line schematic diagram

A different form of one-line schematic diagram is shown in Fig. 9-2. In it the building has three floors and the electric service is assumed to be a 115/230 V, single-phase, three-wire circuit. The source is connected through the kilowatthour meter to the main service-disconnect switch and fuse to the individual circuits included in the panelboard. Because the lighting panels are remote from the main panelboard, a feeder runs from each branch-circuit overcurrent device to its associated light panel.

Figure 9-2 illustrates a major advantage of one-line schematic diagrams. If all parts, such as wires, switches, and fuses were shown pictorially, the diagram would have to be quite large and be very difficult to draw. Also, if three wires were shown for each part of the circuit, the amount of detail would tend to be confusing. Although the specifics conveyed by diagrams of this type are limited, these one-line schematics are ideal for preliminary work.

Final preliminary steps

Once the wiring plans are completed and reviewed, make a careful survey of the building. Pay particular attention to the wall construction in order to ensure that adequate space has been provided for the mounting of outlet boxes. Sometimes the ceilings and walls are furred; that is, thin strips of wood are provided to create an air space or to permit the installation of insulating materials. If, for example, 1″ furring strips are designated for use on a masonry wall, remember that a less than inch-thick mounting board will be available for outlet boxes because the nominal 1″ measurement means that the lumber is really only 3/4″. In such a case, a hole may have to be cut into the masonry wall to make room to sink the outlet boxes. Also, be sure to check on the space available for cables or conduit and note the items which must be avoided—such as water pipes and ductwork for the heating system.

Although a floor plan should show the location of a switch with respect to the walls, it will not show the distance of a switch from ceiling or floor. Only a section (or detail) drawing will give that height, which should then be marked on the drawing to be used during the wiring installation. If the service-entrance location is not shown on available plans, contact the electric supplier since he determines where electric power will enter the building.

Plans are usually drawn to a scale of either 1/8″ or 1/4″ per foot. This scale must be marked on the plans.

9.2   *The Service Drop*

As noted in Sec. 6.1, the service drop is connected and installed by the utility company staff. Although this part of the installation does not involve the nonindustrial electrician directly, it is recommended that you read Sec. 230-24 of the Code to familiarize yourself with the clearance of service drops not exceeding 600 volts. These clearances are maintained as follows: 2 wire, 115 V, single phase; 3 wire, 115/230 V, single phase; 4 wire, 120/208 V, three phase; and 4 wire, 277/488 V, three-phase systems.

Service-drop conductors must, of course, have sufficient ampacity to carry the load. In addition, they must have adequate mechanical strength and must not, in any case, be smaller than AWG #8 copper or AWG #6 aluminum. The only exception to this rule is when the drop supplies only limited loads of a single-phase circuit, such as controlled water heaters. Here, the conductors must be no smaller than AWG #12 hard-drawn copper or equivalent.

9.3   *Service-Entrance Conductors*

As the electrician you are responsible for the installation of the service-entrance conductors. But you must not start the installation until the utility company designates the location of the service entrance. Also, be sure to obtain a copy of the power company's rules, which will be in strict accordance with the local code. Follow them carefully. The service-entrance conductors are usually installed on the outside of the building, but they may be installed on the inside so long as NEC and/or local code requirements are met.

Wiring methods (Sec. 230-F)

Service-entrance conductors extending along the exterior, or entering buildings or other structures may be installed as follows:

(1) As separate conductors, or in cables approved for the purpose, or in cable bus, or enclosed in rigid conduit.

(2) For circuits not exceeding 600 volts the conductors may be installed in EMT, in wireways, in auxiliary gutters, or in busways.

Conductors considered outside building (Sec. 230-44)

Conductors placed under concrete at least 2″ thick or beneath a building, or conductors within a building in conduit or duct and enclosed by concrete or brick not less than 2″ thick are considered to be outside the building.

Mechanical protection (Sec. 230-50)

Individual open conductors or cables other than approved service-entrance cables must never be installed within 8 feet of the ground or where exposed to physical damage. Service-entrance cables must be of the protected type or be protected by conduit, EMT, or other approved means wherever liable to contact with awnings, shutters, swinging signs, or installed in exposed places in driveways, near coal chutes or otherwise exposed to physical damage.

Individual open conductors exposed to weather [Sec. 230-51(c)]

Individual open conductors exposed to weather must be supported on insulators, racks, brackets, or other sustaining means, placed at intervals of not more than 9 feet and so as to separate the conductors at least 6 inches from each other and 2 inches from the surface wired over. Or the supports must be at intervals not exceeding 15 feet if they maintain the conductors at least 12 inches apart. For 300 volts or less, conductors may have a separation of not less than 3 inches where the supports are placed at intervals not exceeding 4-1/2 feet and so as to maintain the conductors at least 2 inches from the surface wired over and with a separation of at least 3 inches between conductors.

Individual conductors entering building or other structure (Sec. 230-52)

Where individual open conductors enter a building or other structure, they must enter through roof bushings or enter through the wall in an upward slant through individual, noncombustible, nonabsorptive, insulating tubes.

U-shaped drip loops must be formed on the conductors before they enter the tubes.

Service cables (Sec. 230-E and F)

Approved service-entrance cables must be supported by straps or other approved means within 12″ of every service head, gooseneck, or connection to a raceway or enclosure and at intervals not exceeding 4-1/2 feet. (The service head, also called *entrance cap* and/or *weather head*, is a fitting at the top of the service conduit that keeps rain from entering the conduit. It consists of three parts—a body which attaches to the service conduit, an insulating block to separate the wires, and a cover to keep the rain out and hold the assembly together.)

Cables that are not approved for mounting in contact with a building or other structure must be mounted on insulating supports installed at intervals not exceeding 15 feet and in a manner that will maintain a clearance of not less than 2 inches from the surface over which they pass.

Service-entrance conductors entering buildings are insulated; those on the exterior of buildings must be either insulated or covered. *Exceptions:* A grounded conductor may be (a) bare copper used in a raceway or part of a service cable assembly, (b) bare copper for direct burial when copper is suitable for soil conditions, (c) bare copper for direct burial without regard to soil conditions where part of an approved cable assembly with a moisture- and fungus-resistant outer covering; (d) aluminum or copper-clad aluminum without individual insulation or covering used in a raceway or for direct burial when part of an approved cable assembly with a moisture- and fungus-resistant outer covering.

To determine which type of service-entrance conductor to use, check with your local electrical inspector or power company. The size of wires to use depends, of course, upon the load which is calculated in the manner demonstrated in Chapter 8 of this book and Chapter 9 of the NEC.

Many times the neutral wire is two AWG numbers smaller than the "hot" wires because it carries less current. With 230 V loads and balanced 115 V loads, the neutral carries no current.

Generally, on an overhead service, service-entrance cables are never spliced except in the meter socket. Exceptions to this rule are given in Sec. 230-46 of the NEC.

Two popular types of service-entrance cable are shown in Fig. 9-3. Notice that both cables have two conductors which are first covered with insulation and then braid. On top of the braid is a bare neutral made from many strands of small wire spun in a flat layer. The cable shown in Fig. 9-3(a) is designated Type SE Style A; the A indicates that the cable is armored. The armor is a galvanized steel tape. The armor is in turn wrapped in rubber tape for protection against moisture. The whole cable is finally jacketed in fabric, usually

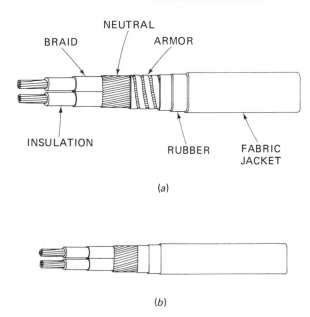

NEUTRAL

BRAID   ARMOR

INSULATION   RUBBER   FABRIC
JACKET

(a)

(b)

Figure 9-3   Two popular forms of service entrance conductors: (a) Type SE, Style A, and
(b) Type SE, Style U

painted gray. The cable shown in Fig. 9-3(b) is designated Type SE Style U; the
U indicates that the cable is unarmored.

9.4   *Service-Head Installation*

After suitable locations have been selected for the service bracket and the
panelboard, the path of the service-entrance cable can be decided. The service
head must be located above the point of attachment of the service-drop
conductors to the building. Where it is impracticable to locate the service head
above the point of attachment, it may be located not further than 24 inches
from the point of attachment.

A weather head with a removable cover is shown in Fig. 9-4. Before
installing it, strip the sheath and the outer insulation from the service-entrance
cable for a distance of 3 feet. If Type SE, Style A cable is being used, remove the
galvanized steel tape for the same distance. Now unwind the fine wires of the
neutral and twist them together to form a single stranded wire. To place the
weather head on the service-entrance cable, first remove the cover and loosen the
cable clamp. Then the weather head can be slipped over the end of the cable
with the three wires extending through the small holes. The cable clamp should
grip the service-entrance cable where the outer insulation and jacket have not
been removed.

Sometimes a special fitting with the shape of a gooseneck is used

Figure 9-4   A typical weatherhead

instead of a weather head. When the fitting is installed it looks like a vertical U. The service-entrance cable is bent to conform to the shape of the fitting and the ends of the cable are taped and painted to seal out moisture.

9.5   *The Meter Socket*

The meter socket or base may be supplied by the power company or by the electrical contractor. In either case, the electrician mounts it 5 to 6 feet above ground level. The meter socket is threaded to accept watertight bushings.

The service-entrance cable is prepared in the manner already described for the weather head, except that less of the overbraid, rubber, tape, and steel tape need to be removed.

The wiring of a meter socket is shown in Fig. 9-5. The service-entrance cable enters the meter socket from the top through a watertight bushing. The watertight bushing clamps the service-entrance cable where its sheath and outer insulation have not been disturbed. The neutral terminal is located in the center of the socket. The two "hot" wires are connected to the input terminals. The service-entrance cable connecting to the panelboard enters the meter socket through a watertight bushing in the bottom of the meter socket. Its neutral wire is also connected to the center terminal and its "hot" wires to the output terminals. Both red wires of the service-entrance cable should be kept on the same side of the meter socket.

In some parts of the country, watthour meters are installed inside buildings. They are usually mounted on the same board as the panelboard. When this mount is used allow space to mount the meter socket above the panelboard. In other parts of the country, a cabinet which is a combination meter socket and panelboard is preferred.

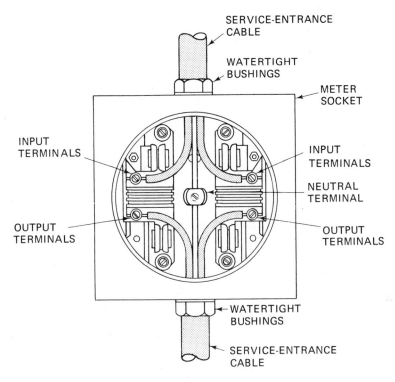

SERVICE-ENTRANCE
CABLE

WATERTIGHT
BUSHINGS

METER
SOCKET

INPUT
TERMINALS

INPUT
TERMINALS

NEUTRAL
TERMINAL

TERMINALS

TERMINALS

WATERTIGHT
BUSHINGS

SERVICE-ENTRANCE
CABLE

Figure 9-5   Meter-socket wiring

9.6   *Entering the Building*

At the point where a service-entrance cable enters a buildings, a *sill plate* is used to keep rain from following the cable into the building interior. A typical sill plate is shown in Fig. 9-6(a). Usually a soft-rubber compound is supplied with the sill plate and is used to seal any opening that might exist.

When the service-entrance conductors are enclosed in conduit, it is customary to use a rain-tight entrance ell at the point of entry into the building.

Figure 9-6   A typical sill plate

With the ell's cover removed, it is a simple matter to pull wires around its right-angle corner.

If the building is of wood-frame construction, you can make the entry hole for cable or conduit by using an electric drill and a hole saw. For large size holes, use an electric saber saw. Use a star drill if you have to cut masonry, such as brick, concrete block, or concrete. If you cut the hole by hand, use a maul or sledge hammer to strike the star drill. Use a power hammer if you can. It considerably reduces the work of drilling masonry.

9.7   *Outside Service-Entrance Miscellany*

Typical fittings used in conduit-type, service-entrance installations are illustrated in Fig. 9-7: a locknut and insulating bushing are shown respectively in (a) and (b): their use is illustrated in (c). The first step is to screw the locknut toward the conduit when starting the locknut. The conduit with the locknut is slipped through the hole in an outlet box or panelboard where the proper size knockout has been removed. If the surface of the panelboard or other device is painted, scrape off the paint so that the locknut will make good contact with the metal. Then screw on the bushing and tighten it with a screwdriver or pliers; finally, tighten the locknut also with a screwdriver or pliers. Because the locknut is inwardly curved next to the outlet box, it will dig into the box. The rounded bushing surface prevents damage to the insulation.

A grounding bushing, Fig. 9-7(d), has a large screw for connecting a grounding wire and a small screw with a sharp point which is tightened after the grounding bushing is firmly in place. The sharp point digs into the metal to ensure a good electrical connection.

In a conduit installation, the weather head, meter socket, entrance ell, and panelboard are joined by pieces of conduit. Nipples (short pieces of threaded rigid conduit) of various lengths up to about 12 inches are sometimes required. Where conduit cannot be run in a staight line—nearly always desirable for appearance sake—prebent elbows are available.

Start the installation by mounting the service bracket and then cutting a hole for the conduit to enter the building. The size of the conduit depends, of course, on the size of the service-entrance conductors. The meter socket must be made for the same size conduit.

Remember that the weather head should be 12″ or more above the service bracket and the meter socket 5 to 6 feet above ground. Assemble the conduit on the ground with the weather head at the top, the meter socket at its correct location, and the entrance ell at the bottom. The entire assembly is often referred to as a stack. A short piece of conduit which joins the entrance ell and panelboard is used as a hinge to swing the stack into position and fasten it to the building by means of approved straps or clamps. The conduit is fastened to the panelboard by using a grounding bushing.

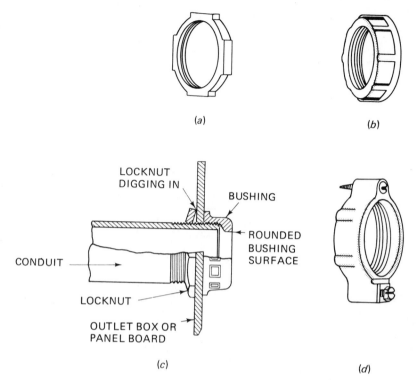

Figure 9-7   Conduit fittings: (a) locknut, (b) bushing, (c) use of locknut and bushing, and (d) grounding bushing

9.8   *Mast-Type Risers*

With ranch-style homes the mast-type riser shown in Fig. 9-8 may be needed to obtain adequate clearance between the service drop and ground. The mast is simply a piece of conduit with a diameter of 2 or 2-1/2 inches to provide sufficient rigidity. A weather head is mounted at the top of the mast. Below the weather head is an insulator for connecting the service-drop neutral. The insulator should be at least 1-1/2 feet above the roof, and the mast should be no more than 4 feet from the edge of the roof. Flashing with an adjustable shield is used at the roof surface to prevent water leaks. Two supports fasten the mast to the building by means of bolts. The fitting at the bottom of the mast is used to connect the conduit that runs to the meter socket.

Wire the mast-type riser the same way you wire a regular conduit-type service entrance.

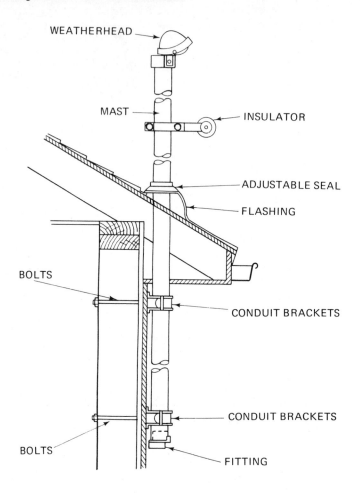

WEATHERHEAD

MAST

INSULATOR

ADJUSTABLE SEAL

FLASHING

BOLTS

CONDUIT BRACKETS

CONDUIT BRACKETS

BOLTS

FITTING

Figure 9-8   Mast-type riser

9.9   *Underground Service*

Underground service is now a common installation method because the availability of relatively inexpensive cable makes the installation economical. In severe winter areas wind and icing problems are eliminated by the use of underground service. Another reason for the increased use of underground services is that many people consider the overhead type unsightly. In some places, all wires and transformers are underground; in others only the wires from the power supplier's pole to the residence are underground.

One form of cable commonly used for underground installation is designated USE (underground service entrance). This cable has an inner layer of

Type RHW insulation that is especially moisture-resistant and an outer layer of insulation that is very sturdy and also water-resistant.

Type UF (underground fused) cable has the Type TW insulated conductors embedded in a moisture-resistant plastic compound that is also very tough mechanically. It may be buried directly in the ground *but it must be protected by fuses or circuit breakers at the starting point.* Type UF cable is generally restricted to farm installations where overcurrent protection is provided at the meter pole which is the starting point. See Article 339 of the NEC for complete details on usage of type UF cable.

Underground cables are buried directly in the ground and no splices are permitted. Both conductors and cables so buried must be at a minimum depth of 18 inches. This depth may be reduced to 12 inches provided supplemental protective covering, such as a 2 inch concrete pad, metal raceway, pipe or other suitable protection is provided.

When a multiconductor cable is installed, the neutral may be bare. When separate conductors are used without conduit, the neutral wire must be insulated like the ungrounded wires. Be sure to properly identify both ends of the neutral so there will be no question when the conductors are connected. Lay the cables in a trench in a loose "snakey" manner. Do not stretch them tightly and do not permit conductors to cross since this may accelerate wear. Individual conductors should be grouped together in a trench.

Metering

A typical outdoor-metering installation for an underground service is illustrated in Fig. 9-9. The meter socket holds the watthour meter and a large-size conduit. The conduit, which projects underground about one-half the depth of the trench, has a conduit bushing on its bottom end. A large multiconductor cable drops out of the conduit with an S-shaped loop rather than in a direct run to the bottom of the trench. The cable loop reduces the chance of cable damage due to earth movement, especially where the earth freezes and thaws.

When the watthour meter is located inside the building, run a piece of conduit through the wall as shown in Fig. 9-10. Be sure to bend the conduit downward. Then seal around both cable and conduit to keep moisture out of the building. The conduit carries the cable directly to the meter socket or to a cabinet which contains both a meter socket and circuit breakers. Although the meter socket may be fastened directly on the wall, the use of a piece of plywood between the wall and socket is preferred. Sometimes the meter socket and panelboard must be mounted a short distance from the place where the underground cable enters the building. In this case, splice the underground cable to the other wires which run to the meter socket. Make splices in a junction box which is later closed with a blank cover.

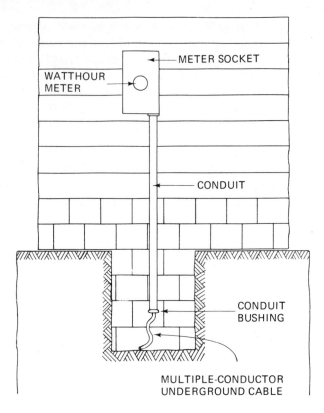

Figure 9-9   Outdoor metering for underground service

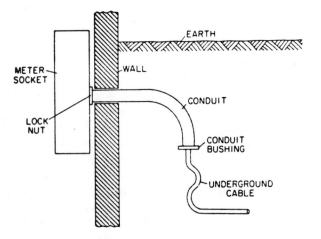

Figure 9-10   Underground cable running through a wall

Pole connections

At the power company's pole, the underground cable must be protected for 8 feet or more above ground. Ordinarily you can do this with a piece of conduit with a weather head on the top and a conduit bushing on the bottom. Use a piece of conduit long enough to extend underground about one-half the depth of the trench. Make an S-shaped loop in the cable where it extends from the conduit to protect it from frost damage.

In some underground electrical services the distribution transformer is placed in a heavy metal vault. You then make connections to the transformer primary inside the vault and connect the service entrance wires in an above-ground splice box. It is also of heavy metal construction.

9.10   *Panelboard Wiring*

Panelboards are made for surface and flush mounting. They have knockouts on top, bottom, sides, and rear. When the panelboards are surface mounted, the rear knockouts are not used since they lay against the wall or a piece of plywood. Panelboards have concentric knockouts so care must be taken in removing these knockouts to keep from making oversized holes. Because the concentric knockouts provide very poor continuity for ground currents, grounding bushings are used with conduit installation.

A typical circuit-breaker type panelboard contains a heavy copper bar called a neutral bus which has numerous terminals for the connection of the systems ground and the neutral wires of all branch circuits. The ungrounded service entrance conductors attach to screw-type terminals embedded in the base of the main circuit breaker. On the load side of the main circuit breaker there is another heavy bus with plug-in provisions for branch circuit breakers.

A wiring diagram is included with each panelboard to help the electrician wire the board correctly.

Run the service-entrance wires into the panelboard and connect them to the appropriate terminals. Be sure to use only Allen wrenches to tighten the Allen-head set screws. They have a six-sided recess rather than a hexagonal head or a screwdriver slot. Place the wires parallel to the sides of the panelboard so that the wiring will have a neat, workmanlike appearance. Avoid sharp bends; use sweeping bends so that the insulation is not pressed against fittings or the panelboard sides. Use raintight fittings and panelboards for exterior installations. When conduit is being installed, use a grounding bushing on the inside of the panelboard. After the set screws are tightened, run wires to the neutral bus or to the separate grounding terminal provided in some panelboards.

9.11   *Grounding*

The usual method of grounding, as noted in an earlier section, is to run the grounding-electrode conductor to the water pipe of the municipal water system. Make sure the water system has metal pipes with conducting fittings. When there is no municipal water system, the steel casing of a well may be used as a ground, but the drop pipe in a dug well can not be used as a ground. Sometimes the metal framework of a building or a gas pipe can be used for grounding. Before using anything other than the municipal water system, however, consult the local inspector and get his approval for the grounding system you propose to use. When the water system is in the ground, the best procedure is probably to run the grounding wire to the nearest cold-water pipe and then place a jumper across the water meter. A jumper is a short piece of wire the same size as the grounding wire. Because a cold-water pipe runs more directly to ground than a hot-water pipe, a cold-water pipe is preferred.

The grounding-electrode conductor may be bare, insulated, or armored. It is always one continuous piece without any splices. A copper grounding wire is never smaller than AWG #8. This size serves the purpose if the largest service-entrance conductors are No. 00 or No. 000. Table 250-94 of the NEC either AWG #1 or 0, use a #6 grounding-electrode wire, and a #4 wire if the service-entrance conductors are No. 00 or No.000. Table 250-94 of the NEC lists the size of grounding-electrode wire to use with larger service-entrance conductors or with aluminum conductors. If it is necessary to use a grounding rod or "made" electrode, the grounding-electrode conductor need never be larger than AWG #6 copper or its equivalent in ampacity.

A grounding-electrode conductor or its enclosure shall be securely fastened to the surface on which it is carried. A No. 4 or larger conductor shall be protected if exposed to severe physical damage. No. 6 wire needs no protection if it follows the surface of the building closely and is securely stapled to it; otherwise, it shall be in conduit, EMT, or cable armor. Grounding conductors smaller than No. 6 shall be in conduit, EMT, or cable armor.

Ground clamps

Select ground clamps carefully, and make sure that the clamp and pipe or rod material are similar. Select a copper or brass clamp, for example, with copper pipe or a copper-coated grounding rod. The ground clamp for armored cable is shown in Fig. 9-11(a). It actually has two clamps, a small clamp for the bare grounding wire and a larger clamp for the armor. When a conduit protects the ground wire, use the ground clamp shown in Fig. 9-11(b). The conduit is grounded by screwing it into the clamp; be sure the other end is grounded at the panelboard. When either bare or insulated wire is used as the grounding wire, use the ground clamp shown in Fig. 9-11(c). Because soldered connections are not permitted in the grounding wire or in the service entrance, all of the clamps shown in Fig. 9-11 make a pressure-type connection. Note also that ground clamps made with a metal strap are also prohibited for use on the grounding wire or in the service entrance.

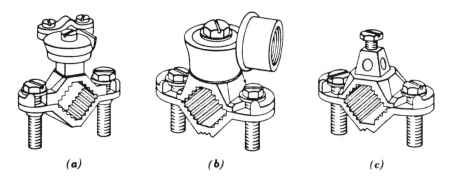

(a)   *(b)*   *(c)*

Figure 9-11   Ground clamps for (a) armored cable, (b) conduit, and (c) bare or insulated wire

EXERCISES

9-1/   In a one-line schematic diagram, what are two common methods of designating the use of three conductors?

9-2/   What is a major advantage of one-line schematic diagrams?

9-3/   To what structural details should you pay particular attention when you survey a building to be wired?

9-4/   What type of drawing will show the height between, say, a switch and the floor?

9-5/   To what scale are wiring plans usually drawn?

9-6/   Normally, what is the smallest size of copper wire that can be used in a service drop? Are there any exceptions to the rule?

9-7/   What materials and tools should you have on hand before attempting to install service-entrance conductors?

9-8/   What methods may be used for installing service conductors extending along the exterior of a building?

9-9/   What conductors are considered to be outside a building?

9-10/   What mechanical protection must be provided for service-entrance cables?

9-11/   How must individual open conductors exposed to the weather be supported?

9-12/   How must approved service-entrance cables be supported?

9-13/   In general, service-entrance conductors entering a building are insulated; those on the exterior of buildings must be insulated or covered. Are there any exceptions to this rule? If so, specify the exceptions.

9-14/   What is the distinction between Type SE Style A and Type SE Style U service-entrance cables?

9-15/   What is a weather head? How is it placed on a service-entrance cable?

9-16/   Where must a service head be located?

9-17/   What is a gooseneck and how is it used?

9-18/   How far above ground should a meter socket be located?

9-19/   Describe how a meter socket is wired.

9-20/   What is a sill plate and how and why is it used?

9-21/   What is an entrance ell and how and why is it used?

9-22/   In a masonry building, describe how you would make the entry hole for service-entrance cable or conduit.

9-23/   How are a locknut and insulating bushing used?

9-24/   In a run of conduit, what is a nipple?

9-25/   How would you assemble and install conduit for service-entrance conductors?

9-26/   Describe a mast-type riser. When and why is it used?

9-27/   What are some advantages associated with an underground service?

9-28/   Describe USE and UF cable.

9-29/   Can Type UF cable be used in the normal service entrance?

9-30/   What is the minimum depth for underground cable?

9-31/   Describe (a) outdoor and (b) indoor metering installations for an underground service.

9-32/   How are pole connections usually made for an underground service?

9-33/   What precautions should be taken in working with knockouts on panelboards? Why?

9-34/   Describe the usual method of grounding systems and several alternative methods which may or may not be permitted by the local code.

9-35/   What is the smallest permissible copper grounding-electrode conductor? What is the limitation for this size of grounding-electrode conductor?

9-36/   For a "made-electrode," what is the largest size copper grounding-electrode conductor required?

9-37/   If a No. 8 copper grounding-electrode conductor is employed, what precautions, if any, must be taken?

9-38/   What precautions, if any, must be taken with a No. 6 copper grounding-electrode conductor?

10
New Work

10.1   *Generalizations*

For you as an electrician, "new work" refers to the installation of wiring systems in buildings under construction. While "old work" obviously refers to wiring installations and/or modifications in buildings where the construction is already completed. In this chapter we are concerned solely with new work; old work will be separately considered in the next chapter of this text.

Roughing in

The installation of conduit or cable, outlet, switch, and junction boxes, and service entrance takes place during the early stages of construction. Collectively, this work is called "roughing in." Wires are pulled into a conduit system only after the lathing, plastering, papering and similar work is completed. Fixtures, switches and receptacles are put in place when the building is nearly completed.

Grounding

Although grounding does not have to be done first, it is convenient to complete this part of the job before device connections are made.

When armored cable is used, the outlet boxes are automatically grounded to the armor. When nonmetallic-sheathed cable is used, there is a choice between cable with or without a grounding wire. Generally, however, cable with a grounding wire is more convenient.

When the wiring is complete, the grounding wire must be continuous from box to box all the way back to the ground terminal (neutral bus) on the panelboard. The removal of any device from a box must be done in such a way that it will not break the continuity of the grounding wire. This rule is needed to assure good grounding and a safe electrical system.

Metallic outlet boxes

It is easy to connect grounding wires properly. One of several different methods is illustrated in Fig. 10-1. The bare grounding wire from each cable is joined to a jumper which connects to the receptacle grounding terminal. A grounding clip is used to fasten the grounding wire to the outlet box. Although a screw may be used in place of the grounding clip, that screw can not also be used for any other purpose, such as clamping the cable. Some outlet boxes are made with a 6″ grounding wire installed and fastened by a screw. Then all you have to do is add a grounding wire for each receptacle and splice all the ground wires together. The splice does not have to be insulated; just make sure you tuck the grounding wires into the bottom of the outlet box carefully. To install a grounding clip, slip it over the end of the grounding wire and then grasp it with a pair of electrician's pliers, as shown in Fig. 10-2. Tap it on until it is tight and

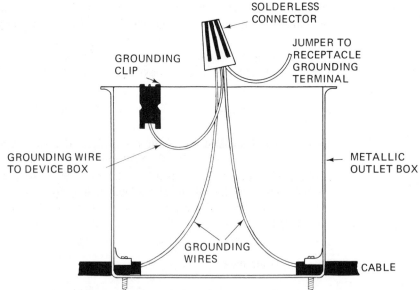

Figure 10-1   Proper connecting of grounding wires

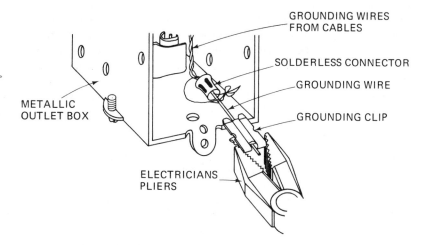

GROUNDING WIRES FROM CABLES

SOLDERLESS CONNECTOR

GROUNDING WIRE

GROUNDING CLIP

METALLIC OUTLET BOX

ELECTRICIANS PLIERS

Figure 10-2   Installing a grounding clip

bend the wire over the edge at the same time. When grounding clips are used, it is best to install them with the grounding wire during rough-in.

When a receptacle is installed directly on a surface-mounted outlet box, the jumper to the receptacle may be omitted because there is direct metal-to-metal contact. When outlet boxes are flush mounted, however, direct contact can not be depended upon unless the receptacles are designed and approved for installation without a jumper. Always use the jumper unless you are sure the receptacle is approved and will be grounded satisfactorily without it.

The grounding terminal on a receptacle is colored green and is hexagonal in shape. Wrap the jumper around the grounding wire and tighten securely.

Nonmetallic outlet boxes

Receptacles installed in nonmetallic outlet boxes must be grounded even though the outlet box itself is not. Use nonmetallic-sheathed cable with a grounding wire. Remember that the grounding wire must be continuous and return without interruption to the panelboard. If the outlet box has a metal faceplate (cover), run the grounding wire to the box and connect it to the cover.

Terminal and lead colors

The terminals and leads on receptacles, lamp holders, lighting fixtures, and other devices are color coded as an aid to correct wiring. Grounding terminals are green, neutral terminals are white, and the terminals for the "hot" wires are natural brass.

The wires on lighting fixtures usually have black and white insulation; alternatively, both wires may be of the same basic color, but one will include a tracer stripe. The wire with the tracer is neutral. If the fixture uses an incandescent lamp, connect the neutral wire to the outer screw shell. When a grounding wire is included, it is either solid green or green with a yellow tracer.

A single-pole switch has two natural-brass terminals. Since switches are used only to interrupt the "hot" side of a circuit, the ungrounded wire attaches to both terminals.

10.2   *Device Locations*

Receptacles

A convenient recommended height for duplex receptacles is 12″ above the floor, because the outlet is readily accessible and the cords of small appliances drag less. In the kitchen, bathroom, laundry, and garage, a good height is 48″ above the floor. Usually, this places the receptacles about 12″ above kitchen-counter tops.

Present practice is to provide receptacles enough so that no point in a room is more than six feet from a receptacle. As a result, the distance between receptacles is less than 12 feet. Any wall space greater than two feet should be counted when receptacles are located. Sliding panels, such as glass doors leading to a patio, are counted as wall space. When sliding doors are used, special receptacles may be required.

Most often, receptacles are spaced equally; however, the spacing is sometimes varied to allow for the probable placement of furniture. A closer spacing of from 4 to 6 feet is recommended above the kitchen-counter top. Also provide a receptacle for the refrigerator and mount it so that it will be hidden when the refrigerator is in place.

To permit the use of power tools and decorative outdoor lighting, install weatherproof receptacles on the exterior house walls at convenient locations. But be sure to keep them at least 18″ above ground level. Special receptacles with key locks are available to prevent vandalism, and for extra convenience, these outdoor receptacles may be controlled by switches mounted inside.

Switches

Switches for outdoor receptacles should be located about 48″ above the floor, on the lock side of the door, and within 6″ of the door frame. Select the switch positions carefully so they are convenient for occupants to use in the normal course of passage from room to room. Upon entering the house, for example, a person should be able to turn on a light while standing in the open doorway, and upon leaving a given room to enter either a hallway or another room, he should be able to light the space entered and turn off the light in the space vacated. This can be accomplished, of course, by using either three-way or four-way switches.

A combination switch and pilot light is often installed when a light, such as one in a basement, can not be seen from the switch location.

In bedrooms where the branch-circuit wires run from the panelboard to a switch and then to a light, consider the use of a combination switch and receptacle. Then the switch is always alive; that is, it controls the light only. This combination is particularly convenient because it provides a ready connection for vacuum cleaners and the like.

Lighting outlets

Wall-mounted lighting fixtures are useful at the sides of a bathroom mirror and should be placed about 4-1/2 to 5-1/2 feet above the floor—depending upon their design and intended use. A fluorescent lamp on each side of a mirror is preferable to either incandescent lamps or a single fluorescent lamp because they provide uniform lighting for make-up, shaving, and other activities requiring unshadowed illumination. Quite often, lamps are mounted on a medicine cabinet which also includes a switch and an outlet, but a separately mounted wall switch is usually more convenient. When an enclosed shower stall is planned, the use of a vapor-proof light fixture, with a wall-mounted switch located outside the shower stall, should also be considered.

The location of special lighting outlets at bookcases, fireplaces, coves, draperies, cornices, and kitchen workspaces calls for special planning. For outdoor use also consider the possibility of installing weatherproof spotlights, floodlights, sidewalk lighting, or post lighting with automatic switching—in addition to any weatherproof receptacles already provided. Remember that for any such exterior lighting control you must use weatherproof switches.

Pendant (suspended) lamps can not be used in closets. The lamp holder can be mounted either on the ceiling or on the wall above the door. When swinging doors are used, the closet lamp may be controlled by a door switch recessed in a jamb. Then, opening the door automatically turns the light on and closing turns it off.

Signal systems

A chime controlled by push-button switches is the most common residential signal system an electrician installs. Systems of this type are supplied from a small transformer which "steps down" 120 V to between 12 and 20 V. Because the transformer power capacity is very limited, it can be connected directly to a 15 A lighting (general purpose) circuit without a special fuse. Annunciator wire is used for connections and should be kept at least 2" from other wires. Many chimes are designed to sound two notes; one for the front door, one for the back door. Push-button switches control chimes and are mounted on the latch side of the door where they are easy to see and push.

In operation, buzzers and doorbells are similar to chimes, but they operate at lower voltages—between 6 and 10 V.

Always take care to make good, low-resistance splices when working with chimes, buzzers, or bells. A high-resistance splice may make the signal system inoperative.

10.3   *Mounting Outlet Boxes*

Outlet boxes are mounted while the wall studs and ceiling joists are exposed. Be sure to learn the thickness and type of the interior wall before you install the boxes. If the wall is noncombustible—made from plaster or gypsum board, for example—the front edge of the outlet box must be no more than 1/4″ behind the final front surface of the wall. If the wall material is combustible—made of wood paneling or the like—the front edge of the outlet box must be flush with the wall.

Outlet boxes are often sold with internal brackets to make their installation easier. The square outlet box shown in Fig. 10-3(a) uses a common bracket. The tabs bent in the bracket help anchor the box. Drive one or two

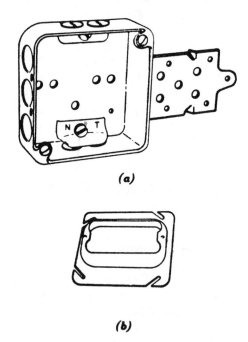

(a)

(b)

Figure 10-3   Square box and raised cover: (a) outlet box with bracket, and (b) cover for switch or duplex receptacle

nails through the small holes in the bracket and into a wooden stud or joist in order to fasten the box securely. Because of their large volume, square outlet boxes also make good junction boxes. They can be adapted for switches or receptacles by using a raised cover like the one in Fig. 10-3(b). Covers are also sold to adapt square boxes to lamp holders. Manufacturers' catalogs show many other cover styles.

A device outlet box with nail holes can be attached to studs. See Fig. 10-4(a). Note that the nails are used outside the box proper. The regular device box with removable sides usually has nail holes in the sides. Use sixteen penny common nails (16d or 3-1/2″ long) to nail the box to a stud. Device-outlet boxes are also made with brackets for easier mounting. Where device-outlet boxes are ganged, the device boxes with mounting ears [Fig. 10-4(b)], can be fastened on metal box supports, as shown in Fig. 10-4(c). Wooden strips at least 1″ thick (nominal lumber size) and mounted between the studs can replace the metal box supports if desired. Small wood screws are driven through the holes in the ears to fasten the outlet box to the wooden strips.

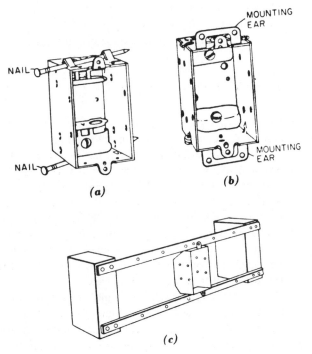

Figure 10-4   Device outlet boxes: (a) with nails, (b) with mounting ears, and (c) on box-support strips

Octagonal outlet boxes are often used with ceiling lighting fixtures and are suspended from ceiling joists by a bar hanger, Fig. 10-5. The center knockout in the bottom of the outlet box is removed so that the stud on the bar hanger can support the outlet box. Adjustable-length bar hangers which fit between studs are also made. Bar hangers are made with various offsets to accomodate different thicknesses of wall material. A 1-1/2" deep box is normally used, but boxes as shallow as 1/2" may be used when it is not practical to use deeper boxes, as may be the case in old work particularly.

Because outlet boxes are made with different cable clamps for nonmetallic cable than for armored cable, be sure you obtain boxes that have matching cable clamps. Of course, if bushings are going to be used, boxes without cable clamps are preferred. The approximate route for cables should be

Figure 10-5   Bar hanger

plotted beforehand so that an outlet box of the proper size can be installed at each location. Remove the knockouts before fastening the outlet boxes in position, but be careful—unused knockouts must be refilled with metal plugs.

10.4   *Cable Installation in Outlet Boxes*

Before inserting armored cable into an outlet box, bend the bonding strip back along the armor. If the outlet box has internal cable clamps, insert the cable into the outlet box where a knockout has been made. Then push the cable into the box until the insulating bushing hits the cable clamp. A correctly installed cable is shown in Fig. 10-6. The cable enters the device box from below. The wires protrude through the hole in the cable clamp which is held in place by a screw. Before you tighten the screw be sure that the insulating bushing or lead sheath is visible and seated against the cable clamp. Also make sure that the bonding strip is placed where it is firmly gripped by the cable clamp when the screw is tightened. Cut off the excess bonding strip. Do not strip the ends of the wires at this time; just tuck them into the box.

Nonmetallic cable is simply slipped into the outlet box until the sheath can be seen in the box. The cable clamp is then tightened and the wires are tucked into the box. The cable clamp shown earlier in Fig. 10-6 is used with armored cable. The cable clamp of Fig. 10-3(a) is used with nonmetallic cable.

When panelboards and outlet boxes do not have cable clamps, use bushings to fasten the cable. Both nonmetallic and armored cables are prepared in the same way. Slip the bushing on the nonmetallic sheathed cable until the sheath shows through the bushing; then tighten the bushing screws. The smaller sized cables have flat sides, so you should rotate the bushing until its flat part mates with the flat sides of the cable before tightening the screws. The bushings

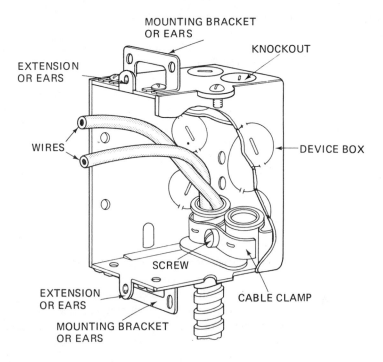

Figure 10-6   Device-box installation

for armored cables have peep holes, so that the insulating bushing or lead sheath can be observed. Remember to place the bonding strip where it will be firmly gripped when the bushing screw is tightened.

Remove the bushing locknut and slip the connector into the knockout of the outlet box or panelboard. The locknut is dish-shaped, not flat; screw it on the bushing so that its teeth dig into the metal. The usual way to tighten a locknut is to use a screwdriver and pliers, as shown in Fig. 10-7. If the cable is armored, the locknut must make good contact with the outlet box, because the bushing and armor form parts of the ground path.

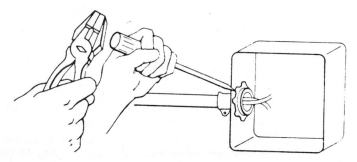

Figure 10-7   Tightening a locknut

10.5 *Wiring Outlets*

The outlets are not wired until the building is nearly completed so the electrician must be careful not to damage or soil wall surfaces while he works. Before you wire the outlet boxes, be sure that they are clean.

When the walls and ceilings are plastered, the outlet boxes are often splattered or even covered with plaster. Plaster may extend over the edges of the outlet boxes and will have to be carefully cut back. In an extreme case, a careless workman may completely cover over an outlet box. To anticipate this, some electricians stuff paper into the outlet boxes after rough wiring in order to keep dirt and plaster out. Clean the boxes thoroughly before you start wiring.

Grounding

There is no special order for connecting the wires in an outlet box, but complete each box before moving on to the next one. Wire one whole circuit at a time rather than sections of several at random. Grounding does not have to be done first, but it is convenient to do it before device connections are made.

In Sec. 10.4 both nonmetallic sheathed cable and armored cable were mentioned without specific consideration of grounding. If armored cable is used, the outlet boxes are automatically grounded by the armor. If nonmetallic sheathed cable is used, choose cable with a grounding wire since all metallic boxes in kitchens, bathrooms, basements, and in any farm buildings, as well as all outdoor boxes must be grounded. Metallic boxes must be grounded if they are in contact with metal lath and are located where either the box or the device it contains can be touched by a person standing on the ground or where a person can touch grounded objects such as plumbing. Standing on a concrete floor is the same as standing on the ground. The grounding wire must be run to every box that contains a receptacle. Other information about the wiring of outlets was covered in Sec. 10.1.

10.6 *Wiring Receptacles*

A duplex receptacle has two white terminal screws on one side and two brass terminal screws on the other side. See Fig. 10-8. A green grounding terminal is on the end. Four terminal screws are provided so that the circuit can be run from one duplex receptacle to the next without splicing.

The duplex receptacle [Fig. 10-8(a)] is shown on a wiring plan by the symbol shown in Fig. 10-8(b). The black ungrounded wire is connected to a brass terminal screw. The white neutral wire is connected to a white screw terminal. Just remember to connect the white wire to a white terminal screw. The remaining wire is bare and is connected to the outlet box by a screw because nonmetallic sheathed cable is being used, as illustrated. When armored cable is

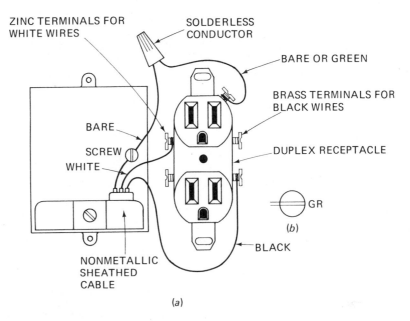

ZINC TERMINALS FOR
WHITE WIRES

SOLDERLESS
CONDUCTOR

BARE OR GREEN

BRASS TERMINALS FOR
BLACK WIRES

BARE

SCREW

WHITE

DUPLEX RECEPTACLE

GR

(b)

BLACK

NONMETALLIC
SHEATHED
CABLE

(a)

Figure 10-8   Wiring a duplex receptacle: (a) wiring diagram, and (b) symbol

used instead, install a short bare wire as a jumper to ground the duplex receptacle to the outlet box.

When terminal screws are used, prepare and wrap the wires as described in Sec. 4.12. Wiring time can be reduced by using copper wire in duplex receptacles equipped with pressure terminals. The terminal colors are molded into the plastic, but the markings vary with the manufacturer. One manufacturer, for example, molds the letters WHITE and GR into the device but does not label the ungrounded terminals.

When only one cable enters an outlet box, as in Fig. 10-8, some of the pressure terminals or screw terminals are unused of course. Often, however, two cables enter an outlet box. In that case, connect one white wire to each white terminal and one black wire to each brass (ungrounded) terminal. Both terminals of a color are common (connected together) so that the circuit is completed through the duplex receptacle. Terminal screws are meant for and approved for securing one wire only. Never place two wires on one terminal screw. When there are more wires than terminal screws, splice the wires and run a jumper wire from each group up to the appropriate terminal.

Two-circuit duplex receptacle

In a two-circuit duplex receptacle each half is independent and has its own terminals. The ordinary duplex receptacle has two terminals on each side, and both terminals on a side are common. Many better duplex receptacles can be converted easily from the ordinary type to the two-circuit type by using a screw driver to remove the contact link which connects the two terminals.

Two-circuit duplex receptacles are widely used where the top half is to be controlled by a switch and the bottom half is permanently connected. Instead of using a permanent ceiling light, for example, a floor or a table lamp may be plugged into the top half and switched. An electric clock, which you want to run continuously, could be switched into the bottom half. Another efficient use of the two-circuit duplex receptacles can be made where two circuits run nearly parallel, as kitchen appliance circuits do. If the two circuits are fed from opposite sides (legs) of the 120/230 V service, only three wires are needed. Then, when two different appliances are plugged into the same duplex receptacle, they will automatically be on opposite legs of the three-wire circuit.

Switched duplex receptacles

The wiring of a switched duplex receptacle is shown in Fig. 10-9. Note

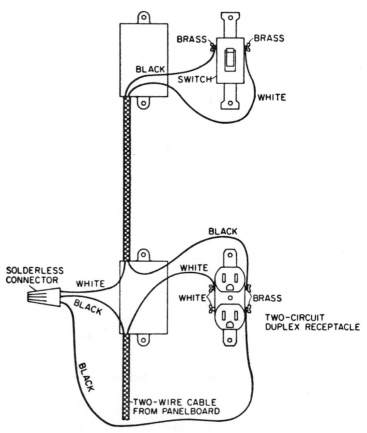

Figure 10-9   Wiring of a switched duplex receptacle

that the terminal screws are shown extended and the grounding wire is not shown—but only in order to make the drawing clearer. Use a convertible duplex receptacle and remove the contact link from the brass (ungrounded) side only. Leave the white (neutral) side connected, and then connect the white wire from the panelboard directly to either white terminal screw and tighten. Splice the black wire from the panelboard to a short piece of black wire and run it to the lower brass terminal on the duplex receptacle. In this way, the lower half is not switched and is, therefore, always ready to supply electric power. Also splice to the black wire from the panelboard a white wire which goes to a terminal on the switch. Then complete the wiring with a black wire from the switch to the top half of the duplex receptacle. The switch now controls the load plugged into the top half of the duplex receptacle.

Previously it was stated that the rule is that the white wire is always reserved for the neutral; yet in Fig. 10-9 we can see that the white wire is spliced to a black wire. Here the white wire becomes an ungrounded wire feeding the switch—and violating the rule. This exception connection is permitted by the NEC in switch loops where two-wire cables are used, because the cables are only made with a black wire and a white wire. Care must be taken to *switch only* the ungrounded wire. Note that black wires are connected to the brass terminal screws on the duplex receptacle and the neutral wire is connected to the white terminal. The wiring would be incorrect if the wires connecting the switch were *reversed* so that a white wire connected to a brass terminal.

Three-wire circuits

The circuits discussed so far have contained two wires; one ungrounded wire and a grounded neutral wire. If two such circuits are run in the same general area, there are four wires, as shown in Fig. 10-10(a). Note that lines *a* and *b* are ungrounded wires and the remaining lines *n* are the neutrals, which are connected to the same neutral bus in the panelboard. If lines *a* and *b* are connected to opposite sides of the service entrance, current will flow as indicated by the arrows. "Opposite sides of the service entrance" means that if

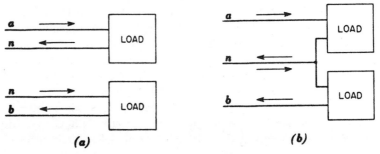

(a)   *(b)*

Figure 10-10   Comparison of (a) two- and (b) three-wire circuits

line *a* is connected to the black service-entrance wire, line *b* is connected to the red service-entrance wire. The voltage between either lines *a* and *n* or *b* and *n* is 120 volts, whereas the voltage between lines *a* and *b* is 240 volts.

Since the neutral wires are connected together at the panelboard, they are, in effect, one wire. Why use two wires when one will do? A three-wire 120/240 V circuit will do the work of two two-wire 120 V circuits. Consider the condition when equal loads are used on the three-wire circuit shown in Fig. 10-10(b). Assume that each load draws 20 A; then 20 A flows in lines *a* and *b*, but the two arrows on the neutral *n* are in opposite directions. As a result, the neutral currents cancel each other, there is no voltage drop in the neutral, and less voltage drop in each circuit. When the currents are unequal, the neutral current is still reduced. If, for example, line *a* is carrying 20 A and line *b* is carrying 15 A, the neutral *n* is carrying 20 − 15 = 5 A. Although the neutral current is no longer zero, it is much less than either current. The single loads in Figs. 10-10(a) and (b) may be replaced by several smaller loads where they are needed if they are about equally distributed between the two sides of the circuit.

A three-wire circuit is wired with a three-wire cable consisting of a white neutral wire, a black wire, and a red wire. Use a convertible duplex socket and remove the connecting link between the brass terminal screws. Wire the white wire to a white terminal screw as with a two-wire circuit. Wire the red wire to one brass terminal screw and the black wire to the other brass terminal screw. *Never wire the red wire to the white terminal screw* or you will have 240 volts across the receptacle. The three-wire cable may be looped from box to box in the same manner as the two-wire cables. Larger outlet boxes may be needed, however, because of the additional wires.

10.7   *Wiring Pull-Switch Lamp Holders*

Pull-switch, or pull-chain lamp holders, are often used in basements, closets, and attics. On a wiring layout a ceiling mounted pull-chain lamp holder is indicated by the symbol shown in Fig. 10-11(a); the wiring of the last one on a circuit is shown in Fig. 10-11(b). The typical lamp holder has one brass terminal screw and one white terminal screw. Connect the black wire to the brass terminal and the white wire to the white terminal.

Not all lamp holders are constructed in the same way, but the wiring does not differ. In some lamp holders the terminals are directly accessible on the top. In a second type, a ring must be unscrewed in order to give access to the threaded area into which the lamp bulb screws. This switch-socket assembly pulls out so that the wires may be placed on the terminal screws. After the wiring is completed, reassemble the lamp holder; the pull switch must be aligned with the insulating base so that the parts slip together. A third type of lamp holder has pressure terminals for use with copper wire. A fourth type is equipped with 6″ leads; splice the black lead to the black wire and the white lead to the white wire.

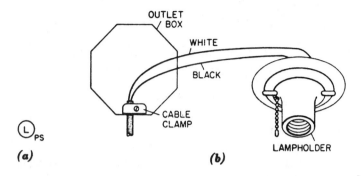

Figure 10-11   Pull-chain switch lamp holder wiring: (a) symbol, and (b) wiring diagram

It is usually necessary to splice wires that continue on to another outlet because most lamp holders have only two terminal screws. With receptacles, splicing is unnecessary because they have four terminal screws. When the cables continue to another outlet, the outlet is wired by using two short pieces of wire (jumpers) as shown in Fig. 10-12. Each jumper should be about 6″ long and is made usually from a piece of wire of the same size as that used in the cable. Connect the black jumper to the brass terminal of the lamp holder and the white jumper to the two black wires from the cables and splice the white jumper to the white tires from the cables.

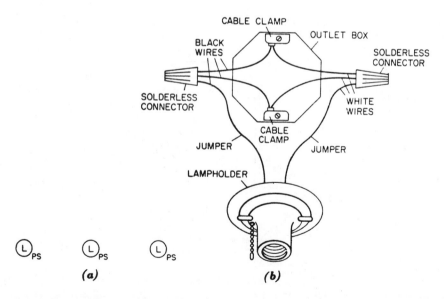

Figure 10-12   Wiring of a pull-chain switch lamp holder when cable continues to another outlet: (a) symbol, and (b) wiring diagram

Switched lamp holder

A lamp holder or a lighting fixture controlled by a single-pole switch is shown by the symbols in Figs. 10-13(a) and (b). A dash below the symbol S in Fig. 10-13(a) shows that the cable from the panelboard runs to the switch. If there is no dash below the switch symbol, the cable from the panelboard runs to the lamp holder. When the cable runs to the switch first, the wiring is done as shown in Fig. 10-13(c). At the device box, the two white wires are spliced together. Each black wire is connected to one side of a toggle switch. At the outlet box the black wire is wired to the brass terminal screw of the lamp holder and the white wire is wired to its white terminal screw.

When the cable from the panelboard runs directly to the lamp holder outlet box, the connections are made as shown in Fig. 10-13(d). The white wire from the panelboard is wired directly to the lamp holder white terminal, but the black wire is spliced to the white wire in the cable running to the switch. The black wire from the switch is connected to the brass terminal on the lamp holder. At the switch the white and black wires are each connected to a switch terminal.

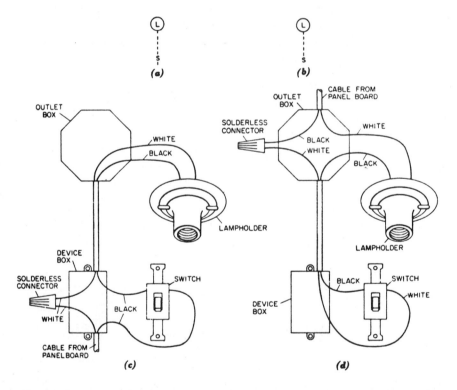

Figure 10-13   Wiring of switched lamp holders: (a) symbol for switch feed, (b) symbol for lamp holder feed, (c) wiring for switch feed, and (d) wiring for lamp holder feed

In both Figs. 10-13(c) and (d) the lamp holder is connected to one black wire and one white wire. Note, however, the difference in the switch wiring. Both wirings are correct and both are approved by the NEC. In Fig. 10-13(c) the switch is connected by two black wires, but in (d) the switch is connected by one black wire and one white wire. This change in wire colors is permitted when two- or three-wire cables are used. When conduit is used, the white wire from the outlet box to the device box would be replaced by a black wire. In both Figs. 10-13(b) and (d) only the ungrounded wire is switched; the neutral is not broken by a switch, although it is spliced in (c).

Although a lamp holder is shown in both Figs. 10-13(c) and (d), a lighting fixture could be used just as well. An incandescent lighting fixture usually would have one black wire and one white wire. Some fluorescent lighting fixtures have one black wire and two white wires. Just splice all the black wires in one group and all the white wires in another group.

10.8   Two Lamp Holders Controlled by One Switch

Often two or more lighting fixtures are controlled by a single switch. The symbols in Fig. 10-14(a) show two lamps controlled by a single-pole switch.

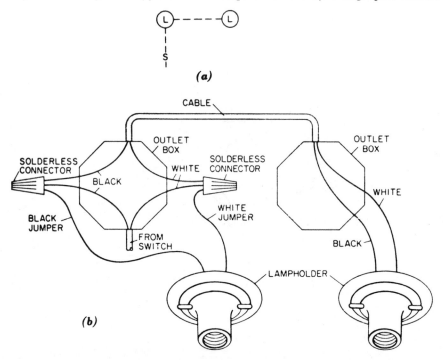

Figure 10-14   Wiring for two lamps controlled by one switch and using switch feed: (a) symbol, and (b) wiring diagram

Switch feed is used and the switch is wired the same as in Fig. 10-13(c). Again, the white wire from the panelboard is spliced to the white wire feeding the switch. The wiring of the right-hand outlet box is the same as in Fig. 10-13(c). At the left-hand outlet box a cable and two jumpers are used. All the black wires have been spliced together in one connector and all the white wires together in another. The black jumper is wired to the brass terminal and the white jumper is wired to the white terminal.

When the cable from the panelboard runs to a lighting fixture instead of to a switch, the wiring is slightly more complicated. The symbols for such a circuit are shown in Fig. 10-15(a) and the wiring is shown in Fig. 10-15(b). One cable and two jumpers have been added to the wiring used for one lamp holder. Again the black wire from the panelboard is spliced to the other black wires. The white wire from the panelboard is spliced to the white jumper and to the white wire of the cable feeding the second lighting fixture.

In these instructions one cable is always specified as coming from the panelboard. In practice, the cable does not usually run directly from the panelboard but instead runs from another outlet box where power is always

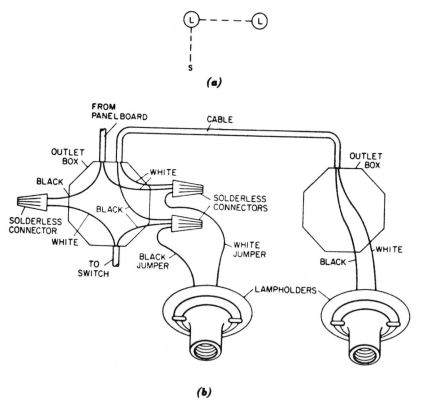

Figure 10-15   Wiring for two lamps controlled by one switch and using lamp holder feed: (a) symbol, and (b) wiring

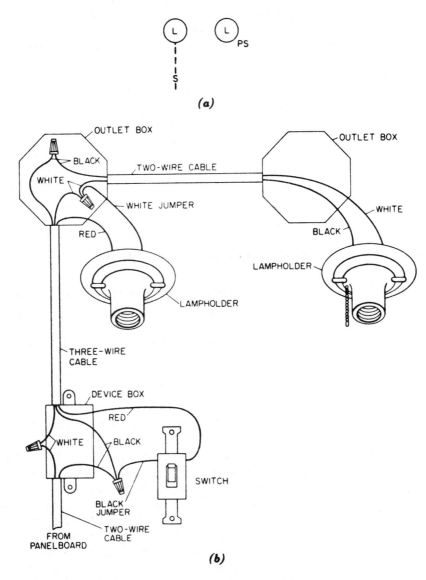

Figure 10-16   Lamp holder wiring using three-wire cable: (a) symbols, and (b) wiring

available. For example, several lamp holders are usually wired on the same circuit. The cable runs from lamp holder to lamp holder, and only the first lamp holder is wired directly to the panelboard. An example of this is shown in Fig. 10-16, where a pull-switch lamp holder is fed from the switch that controls a switched lamp holder. This diagram also shows one place where three-wire cable can be used to reduce the amount of wire in a residence.

Although the symbols in Fig. 10-16(a) show no connection between the two lamp holders, some of the wiring may be shared as in Fig. 10-16(b). The device box is fed by a two-wire cable, but a three-wire cable runs to the outlet box for the switched lamp holder. Another two-wire cable runs to the pull-switch lamp holder. At the device box the two white wires are spliced and run to the switched lamp holder, where a white jumper is spliced to the white cable wires. Although it is spliced, the white wire is continuous to the pull-switch lamp holder. A black jumper is used on the device box to connect the switch. The black wires are spliced here and in the first outlet box to make a continuous wire to the pull-switch lamp holder. The third (red) wire of the three-wire cable is connected to a switch terminal and to the switched lamp holder. The red wire is, of course, wired to the brass terminals on the switched lamp holder.

10.9   *Switches With Pilot Lights*

Wiring a switch pilot light is similar to wiring a switch to control two lamps at the same time. The difference is that one lamp is mounted at the switch instead of both lamps at a distance from the switch. The symbols for the wiring are shown in Fig. 10-17(a), where the cable feeds the switch, and in Fig. 10-17(b) where the cable feeds the lamp holder. In Figs. 10-17(c) and (d), the combinations of switch and pilot lamp are shown with the switch and pilot lamp each having independent terminals. Some manufacturers' products may have an internal lead between the switch and the pilot lamp and will require slightly different wiring.

When the cable from the panelboard feeds the switch, as in Fig. 10-17(c), only two-wire cables are needed. The white wires are spliced in the device box with a white jumper, and the jumper is connected to the white terminal on the pilot lamp. The black wire from the panelboard is connected to one side of the switch, and the black wire running to the lampholder is connected to the unused switch and pilot lamp terminals. Because there usually will be a barrier between the terminals, strip this wire as shown in Fig. 10-18 to eliminate making a splice. The bare wire between the insulation may be wrapped around one terminal screw and the bare wire on the other end wrapped around the other screw terminal.

When the cable from the panelboard feeds the lamp holder, as in Fig. 10-17(d), it is convenient to use a three-wire cable between the outlet box and the device box. The white wire connects to the white terminals of the lamp holder and the pilot lamp. The black wires from the two cables are spliced in the outlet box. At the device box the black wire is connected to a switch terminal, and at the outlet box the red wire is connected to the lamp holder brass terminal. At the device box the red wire is stripped as shown in Fig. 10-18 and connected to the brass terminal of the pilot lamp and a switch terminal as shown in Fig. 10-17(d).

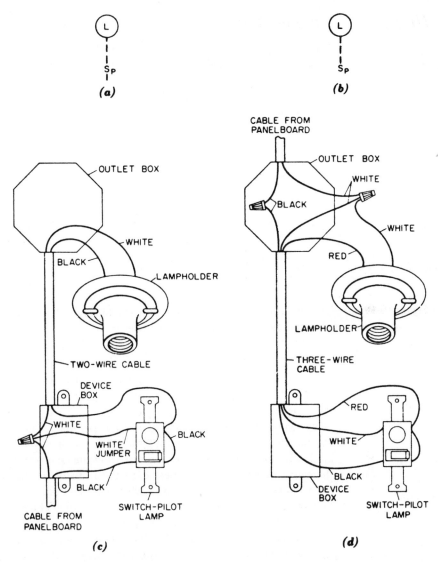

Figure 10-17 Wiring of switch with pilot light: (a) symbol for switch feed, (b) symbol for lamp holder feed, (c) wiring for switch feed, and (d) wiring for lamp holder feed

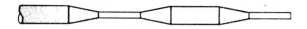

Figure 10-18 Wire stripped for combination switch-pilot lamp

10.10   *Three-Way Switches*

When a pair of three-way switches control a single outlet, there are several combinations in which the cable from the panelboard, the switches, and the outlet may be arranged. The outlet may be either a lamp holder or a receptacle. The most common connections are shown symbolically in Figs. 10-19(a), (b), and (c). In Figs. 10-19(a) and (b) the cable from the panelboard feeds the lamp holder. In Fig. 10-19(c) the cable from the panelboard feeds a three-way switch.

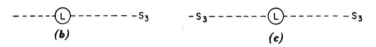

(a)

(b)   (c)

Figure 10-19   Symbols showing combinations for three-way switch connections: (a) and (b) lamp holder feed, and (c) switch feed

In three-way switches the common terminal is a different color from the other two and its location varies with the make of switch. In the following examples the common terminal is shown alone on one side. The white wire from the panelboard will always run to the white terminal of the lamp holder without being switched, although it may be spliced. The black wire from the panelboard will be connected to the common terminal of one three-way switch with either a black wire or a white wire and the wire from the common terminal of the other three-way switch will be connected to the lamp holder brass terminal. At the lamp holder there will always be a white wire and one of another color. When three-wire cable is used, the other wire will be colored either black or red.

The wiring for the three-way switch circuit of Fig. 10-19(a) is shown in Fig. 10-20(a). This is the easiest of the circuits to wire because the fewest splices are made. The white wire from the panelboard is wired directly to the lamp holder, but the black wire is spliced to a white wire from the common terminal of switch S_2. In the device box for switch S_1 the white wires from the two-wire and the three-wire cables are spliced. At both switches S_1 and S_2 the red wires are connected to the upper terminals and the black wires to the lower terminals on the same side of the switches. Either or both of these two wires from the three-wire cable could be reversed and the circuit would still function properly. The black wire on the outlet box at the lamp holder is part of the two-wire cable and is connected to the common terminal of S_1.

The wiring for the circuit of Fig. 10-19(b) is shown in Fig. 10-20(b). A two-wire cable from the panelboard supplies power to the lamp holder outlet

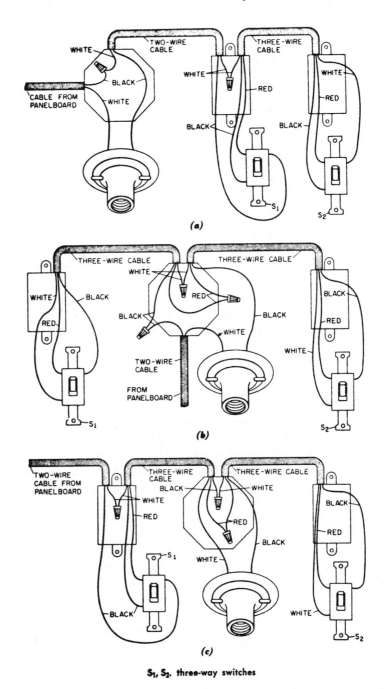

S₁, S₂. three-way switches

Figure 10-20   Three-way switch wiring: (a) Lamp holder feed with switches on one side, (b) lamp holder feed with lamp holder in the middle, and (c) switch feed

box, and two three-wire cables connect to two device boxes for switches. Again the white wire from the panelboard is connected to the white terminal of the lamp holder. This time the black wire from the panelboard cable is spliced to the black wire of the cable running to S_1, where it is connected to the common terminal of S_1. At both S_1 and S_2 red wires are connected to the upper terminals and white wires to the lower terminals on one side of each switch. At the outlet box for the lamp holder both white wires from the three-wire cables are spliced. The red wires are also spliced in this outlet box. The common terminal of S_2 is connected to the lamp holder by a black wire.

The wiring for the circuit of Fig. 10-19(c) is shown in Fig. 10-20(c). A two-wire cable from the panelboard supplies the power, and three-wire cables connect the boxes together. This time the white wire from the panelboard cable is spliced in the device box for S_1 to a white wire from the three-wire cable. At the lamp holder outlet box this white wire is connected to the lamp holder white terminal. The black wire from the panelboard cable is connected to the common terminal of switch S_1. In the outlet box in the center the red wires from S_1 and S_2 are spliced together. In the same outlet box, the black wire from S_1 is spliced to a white wire from S_2. The brass terminal of the lamp holder is connected to the common terminal S_2 by a black wire. Note that S_1 is connected by a red wire and two black wires, whereas S_2 are connected by a red wire, a black wire, and a white wire.

10.11   Door Chimes and Door Bells

Transformers for use with door chimes and door bells are made with fixed outputs of either 16 or 10 volts. The usual mounting methods are through a 1/2″ knockout of an outlet box or on a flat plate which covers an outlet box. With these mountings the transformer is outside the outlet box, the primary wires are inside the outlet box, and the connecting screws or wires for the low-voltage (secondary) winding are outside the outlet box. Transformers are also made for surface mounting inside an outlet box or panelboard, but they take up space which could be used for making other connections.

The primary of the transformer is connected to the wires of a 120 V lighting or special-purpose circuit by splicing within an outlet box. One primary wire is connected to the neutral wire, and the other wire is connected to an ungrounded wire. The secondary connections are made with bell wire. The secondary connections for door chimes are shown in block diagram form in Figs. 10-21(a) and (b). In Fig. 10-21(a) the door chime can be wired to sound a double note for the front door and a single note for the back door. When wiring the two-door chime, run one wire from one transformer secondary terminal to the terminal labeled TRANS of the chime. Run another wire from the transformer secondary terminal to one terminal on the front-door switch and one terminal on the back-door switch. The wire may be spliced at a convenient place rather than run from the front-door switch to the back-door switch. Run a wire from the unused terminal on the front door switch to the FRONT terminal of the chime. Run a wire from the unused terminal of the back-door switch to the REAR (BACK) terminal of the chime. When the two-door chime is tested it should sound the proper notes for both doors.

A one-door chime is often used with only one door switch, but the connections for using a one-door chime with two door switches is shown in Fig. 10-21(b). The wiring is very similar to that for the two-door chime except that the one-door chime has only two terminals; so the wires from the front- and back-door switches are spliced together and a single wire is run to the chime. If only one switch is to be used, the connections to the second switch are not installed. Then, for example, one wire will run from the front-door switch to the transformer and another wire from the same switch to the chime terminal. Additional switches can be wired in parallel to the switches shown, or the one-door chime can be replaced by a bell or a buzzer.

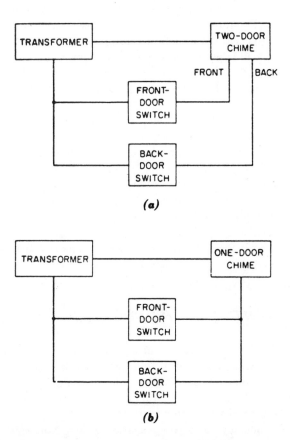

Figure 10-21   Door-chime connections: (a) two-door and (b) one-door chime

10.12   *Appliance Circuits*

The NEC classifies appliances in three groups: fixed, portable, and stationary. Fixed appliances are permanently fastened in one place; for example, water pumps and heating-system motors. Portable appliances are those which can be easily moved around in normal use. Stationary appliances usually remain where they are installed, although they are not fastened into place. For instance, a stationary appliance like a refrigerator can easily be moved from one house to another.

Every appliance must be provided with a disconnect means and overcurrent protection. With portable appliances the plug and receptacle arrangement is satisfactory if the plug and receptacle ratings are equal to or greater than the appliance ratings.

Fixed and stationary appliances rated below 300 W or 1/8 hp require no special disconnecting means. The branch circuit fuse or circuit breaker provides adequate protection.

When larger fixed and stationary appliances are connected to a circuit serving other loads, no special disconnect means is needed if the circuit is protected by a circuit breaker or fuses mounted on a pull-out block. If the circuit is protected by plug fuses, however, you must install a separate enclosed switch for each appliance. Although the switch does not have to be fused, the usual practice is to install a fused switch. Enclosed switches contain one fuse for 120 V circuits and two fuses for 240 V circuits. Install fuses which have an ampere rating not greater than that of the fuse protecting the branch circuit.

When a fixed or stationary appliance is supplied by its own individual circuit, follow the above recommendations. However, the fuse or circuit breaker ratings should not exceed 150 percent of the appliance ampere rating unless the appliance is rated at less than 10 A. In that case, the 15 A fuse or circuit breaker may be used. If the appliance has an automatically started motor, the overcurrent device should not be rated over 125 percent of the ampere rating of the motor being protected. Although the NEC does not require a separate circuit for each automatically started motor, such as a water pump or oil burner, it is wise to provide separate circuits.

10.13   *Wiring for Appliances*

Smaller appliances can be connected in the standard receptacle, which is rated at 15 A. Heavy appliances often need larger receptacles with special contact arrangements, such as those shown in Figs. 10-22(a) and (b). The receptacle in Fig. 10-22(a) is rated at 30 A and 250 V and is commonly used for electric clothes dryers. The receptacle in Fig. 10-22(b) is rated at 50 A and 250 V and is commonly used for electric ranges. Note the difference in the contact arrangement which prevents the wrong plug from being inserted. These receptacles may be used in either three-wire 120/240 V circuits or 240 V circuits, with the third contact used for grounding.

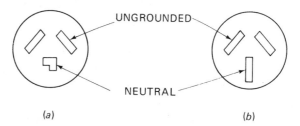

Figure 10-22 Contact arrangements for common power receptacles: (a) clothes-dryer receptacle, and (b) range receptacle

Some 240 V appliances, such as water heaters, do not need connections to the neutral and may be wired by a two-wire cable. However, the standard two-wire cable is made up only with black and white wires. The white wire should be used only for the neutral, which creates a recognition conflict. However, if both ends of the white wire are painted black, the wire can be considered black and then be used as an ungrounded wire. Some other appliances, such as electric clothes dryers, may operate on a 120/240 V combination and do require a neutral wire.

Heavy appliances may be wired with either cable or conduit. Ranges and clothes dryers may be wired with service-entrance cable, which has a bare neutral. When the appliance is to be connected by a cord and a plug, run the cable up to the receptacle which may be either flush or surface mounted. Flush-mounted receptacles are used with square outlet boxes and wall plates. Follow the receptacle manufacturer's recommendations when deciding which size of outlet box to install. Locate the receptacle within 6 feet of the intended appliance location. Good planning makes it possible to locate the receptacle closer to the appliance and thus simplify the final installation. The cords, or pigtails, for electric ranges and clothes dryers are made in lengths up to 6 feet, but these appliances are more likely to be delivered with cords about 2 feet long.

Clothes dryers are basically 240 V loads, although some have 120 V motors. Clothes dryer frames must be grounded. You can do the grounding with the neutral wire if it is No. 10 or heavier. Thus a three-wire cable must be used even if the entire clothes dryer operates at 240 V. The 30 A receptacle in Fig. 10-22(a) and the No. 10 wires are heavy enough for the typical home clothes dryer. Run the white wire to the neutral terminal and each ungrounded wire to one of the ungrounded terminals. Tighten the connections securely, because high currents will be carried. Although the NEC permits the use of the plug and receptacle device as the disconnecting means, a separate switch may be installed when the circuit is protected by plug fuses.

A water heater which is rated at less than 4600 W may be supplied by a two-wire No. 12 cable. Paint both ends of the white wire black and protect the installation with a 20 A two-pole circuit breaker, or fuses on a pull-out block, or a separate fused switch. Although grounding is not required, connecting a ground wire to a water pipe is recommended. Sometimes a plastic pipe is used for plumbing or the metal water pipes may be disconnected; therefore, a grounding wire makes the water heater safer. If the branch circuit is protected by plug fuses, install a separate switch. If the branch circuit is protected by a two-pole circuit breaker or fuses on a pull-out block, a separate switch is not needed.

10.14   *Ranges*

The wiring for self-contained ranges is similar to that for electric clothes dryers. Because of the manner in which the oven and individual burners are connected within the range, the neutral wire carries less current than the ungrounded wires. So you may use a neutral one size smaller than the ungrounded wires. For most ranges, the ungrounded wires are No. 8 and the neutral is No. 6. If the range is connected by a pigtail cord, use a receptacle which has the contact configuration shown in Fig. 10-22(b). Connect the neutral wire to the neutral terminal and the ungrounded wires to the ungrounded terminals. The pigtail cord serves as the disconnecting means and serves to ground the range through the neutral wire.

In sectional ranges the oven is a separate unit installed in the wall or between kitchen cabinets, and the burners are installed on the kitchen counter at a convenient location. The NEC calls the units wall-mounted ovens and counter-mounted cooking units. Self-contained ranges are considered stationary appliances while such sectional ranges are considered fixed appliances.

The two basic methods of wiring sectional ranges are either to supply each section by its individual branch circuit or to supply both sections from one 50 A circuit. Any wiring method may be used, including the use of service-entrance cable with a bare neutral. The frames of the sectional units may be grounded to the neutral wire if the wire is No. 10 or larger. If the sections being installed do not have the frames connected to the neutral terminals, make sure they are grounded.

Wall-mounted ovens are usually rated for about 4600 W. At 240 V the current is nearly 20 A, so No. 12 would be suitable. Use No. 10 for grounding. The wires may run directly to the oven, but it is more convenient to install the same kind of 30 A pigtail cord and receptacle that is recommended for clothes dryers. A counter-mounted cooking unit is installed in the same manner as the wall-mounted oven. Use No. 10 wire unless the section is rated above 6900 W; then it requires heavier wires.

The wall-mounted oven and the counter-mounted cooking unit may be supplied by a single 50 A cable containing two No. 6 wires and one No. 8 wire. The circuit for this wiring is shown in block-diagram form in Fig. 10-23. A 50 A

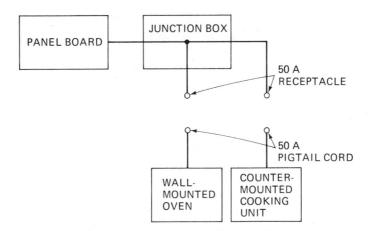

Figure 10-23   Sectional range supplied by a 50 A cable

cable is run from the panelboard to the junction box where it is tapped to make the two cables connecting the 50 A receptacles. Below the receptacles the 50 A pigtail cords are used to connect the wall-mounted oven and the counter-mounted cooking unit. If two junction boxes are mounted as close as possible to the section and only enough wire to permit servicing is used, No. 10 wire can be run from each junction box to each section. Receptacles and pigtail cords are not required, but they will make the installation more convenient. Alternatively, the sections may be wired directly to the junction box shown on the branch circuit.

Pigtail cords are not considered as disconnecting means for fixed appliances. The disconnecting means may be a two-pole circuit breaker, cartridge fuses on a pull-out block, or a separate enclosed switch with cartridge fuses.

10.15   *Electric Space Heaters*

Fixed electric space heaters are supplied from individual branch circuits rated at 15, 20, or 30 A. Because the heaters are considered to be continuous loads, the circuit current can not be over 80 percent of the wire ampacity. For example, a 20 A branch circuit should not carry over $0.80 \times 20 = 16$ A. When a long cable supplies the heater, a larger size cable may be needed to reduce the voltage drop and maintain the heater output.

A baseboard-type heater controlled by a two-pole line-voltage thermostat is shown in Fig. 10-24(a). The heater is supplied by a two-wire 240 V circuit with an ampacity of 20 A. The typical two-pole line-voltage thermostat is rated at 5000 W, but the load should be limited to 80 percent of the rating, or 4000 W. Because a No. 12 copper wire rated at 20 A is used at a reduced ampacity of 16 A, the two-pole line-voltage thermostat will not normally control more than $16 \times 230 = 3680$ W. Although the standard circuit voltages are 120 and 240 V, all calculations are made with 115 and 230 V to agree with the NEC.

Thermostats are mounted on $2'' \times 3''$ device boxes. In Fig. 10-24(a), a two-wire cable from the panelboard runs to the two-pole line-voltage thermostat. Another two-wire cable runs from the two-pole line-voltage thermostat to the overheat switch on the heater. The overheat switch is a thermostat which runs the full length of the heater; and it shuts the heater off if its temperature rises too high. This could happen, for example, if the heater were completely blocked off by draperies. The overheat switch is made part of the heater to increase the safety of the installation. From the overheat switch a wire runs to one connection of the heating element. The other wire from the two-pole line-voltage thermostat connects to the other terminal wire of the heating element within the heater. When the two-wire cable contains a grounding wire, the grounding wire is fastened to the heater frame.

The low-voltage control system in Fig. 10-24(b) uses a thermal relay to control the power to the heater. The thermal relay includes a transformer T,

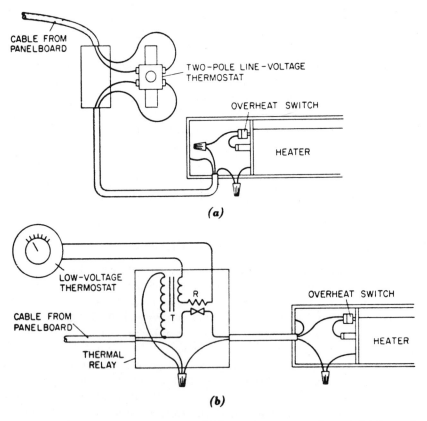

Figure 10-24   Wiring of electric space heaters: (a) line-voltage control, and (b) low-voltage control

which steps 240 V down to 24 V. The low-voltage thermostat controls the flow of current through a resistance heater R. When the thermostat closes, calling for heat, a current flows through resistance heater R, which is wound on a bimetallic strip made from two suitable different metals. The resistance heater heats the bimetallic strip and causes it to bend and close the switch contacts of the thermal relay. When the room temperature rises, the thermostat opens and stops the flow of current through the resistance heater, so that the bimetallic strip cools. After the bimetallic strip cools, the switch contacts of the thermal relay open. In this way the thermal relay silently controls the power flowing to the heater.

In Fig. 10-24(b) the cable from the panelboard runs to the thermal relay, where the transformer primary is connected between the two wires. After the splice, one wire runs directly to the heating element, but the other is connected to a terminal on the switch of the thermal relay. From the thermal relay terminal a wire runs to the overheat switch of the heater and then to the heating element.

In Fig. 10-24(b), the diagram shows how the thermal relay directly controls the power flow to the heater. The thermal relay could also be used to control an electrically operated switch such as a relay or contactor. This flexibility permits you to place the low-voltage thermostat and thermal relay in the most convenient places while running short cables to the heater.

Fixed electric space heaters are rated as fixed appliances and should be protected as explained previously.

10.16   *Testing*

Continuity tests

Branch circuits are sometimes tested for short circuits and continuity with a tester made from a door bell or a buzzer in series with a battery consisting of two 1.5 V cells. The test is made after the wiring is completed but before the power is connected. Door bells, water heaters, and appliances with thermostats and automatic switches must be disconnected.

Branch circuits may be quickly checked for short circuits by connecting the free lead of the tester to the neutral bus of the panelboard. When the test lead from the battery is touched to any ungrounded wire, the tester should not ring. If the tester rings, it indicates a short circuit either between the ungrounded wire and the neutral or between the ungrounded wire and the metallic outlet box.

If the test does not indicate any short circuits, the continuity test can be made. Turn all circuit breakers on and install all fuses. Connect temporary jumpers between the neutral bus and both input terminals of the panelboard. Be sure to remove the jumpers after the test is completed.

Then take the tester to each outlet where a fixture or receptacle is to be installed. Touch it across the black and white wires; it should ring to indicate continuity. The circuit should be completed through the tester just as it will be completed later through a lamp or an appliance. To make the test, you must short together the wires which will run to switches in order to duplicate conditions when switches are turned on. Because cables are looped from outlet box to outlet box, you will need temporary connections with duplicate splices in outlet boxes and at receptacles. After this test, connect the tester between the black wire and the metal box. If the grounding wire or armored cable is well-grounded at each box, the bell will ring. Because the grounding circuit has a higher resistance than the neutral wire, the bell may not ring loud. The grounding should be well done as a safety precaution.

Insulation test

The NEC requires that all wiring shall be so installed that the system will be free from short circuits and unspecified grounds. Most such faults are caused by faulty insulation. Determine whether the insulation of the wiring has been damaged by crushing, by excessively sharp bends, or by penetration by nails. Sometimes switches, receptacles and other devices may be poorly insulated because of in-transit breakage, or defective materials, or improper installation. The wiring testing can be best done after all switches, receptacles, and circuit breakers are installed but before the lamps are installed and the appliances and other loads are connected.

The test door bell will detect short circuits but will not detect damaged insulation when the resistance is high: 100 ohms, for example. The resistance of insulation can be better measured with either a VOM (volt-ohm-milliammeter) or a Megger. (A VOM can also be used, of course, to measure voltages, resistances, and currents.) Insulation is measured by placing the instrument switches in the proper positions to measure the highest resistance. The test leads are then touched together for a moment. The resulting short circuit should be indicated on the instrument scale as zero ohms. If the pointer does not indicate exactly zero, correct the pointer position with the zero-ohm adjusting knob. This procedure is called zero adjustment and should be carried out before short-circuit tests. The test leads are then connected to the circuit under test and the resistance is read on the proper scale of the instrument.

A Megger is a measuring instrument that is made specifically to measure high resistances; it is superior to the VOM because it tests with a high voltage, which makes it possible for it to find defects missed by the VOM. Meggers usually operate at voltages between 150 and 500 V, whereas the VOM is usually limited to about 7.5 V. The leads of a Megger are connected to the circuit to be tested and the resistance is read directly in megohms.

Common wiring faults

Three common faults of electrical wiring are open circuits, short-circuits, and grounds. An open circuit, as the name implies, is simply a circuit that is no longer complete because the continuous path for the current has been broken accidentally. As might be expected, the failure occurs largely at splices and at terminals where the mechanical fasteners may loosen and open the circuit. The danger from this type of fault is arcing at the point where the circuit opens. This arcing can cause considerable damage without tripping the circuit breaker or blowing the fuse unless a ground or short-circuit results from the original fault.

A short-circuit results when there is contact between the ungrounded and neutral wires which bypasses the load. Because the resistance of the circuit is considerably reduced during the short-circuit, a high current flows in that part of the circuit. The high current trips the circuit breaker or blows the fuse. The direct result is a dangerously high current flow which overheats the insulation and may even cause fire.

A ground, or a ground fault, in a system occurs when an ungrounded wire accidentally comes into contact with the ground or the grounding wire. Since the resistance of the ground can be quite high, 20 ohms or more, a ground fault may not trip the breaker or blow the fuse. The first indication of the ground fault may come as a high electric bill or some difficulty in operating the appliances on the grounded circuit. A grounded circuit always presents a possible shock hazard. This is illustrated in Fig. 10-25. The post light is mounted on the

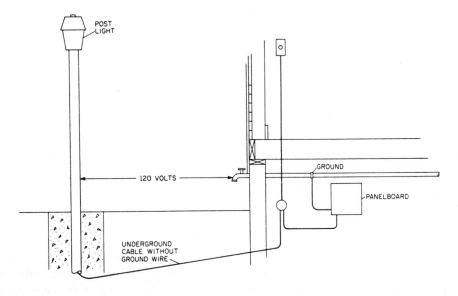

Figure 10-25   Ground fault in post light

upper end of a metal pipe which has its lower end buried in concrete. Because the post light is fed by a two-wire cable, it is not connected directly by a wire to the ground of the electrical system. There is a ground path through the earth, but its resistance is high compared with the resistance of a copper grounding wire. Therefore, when the ungrounded wire accidentally touches the metal pipe, not enough current flows to trip the circuit breaker or blow the fuse. Consequently, 120 volts exists between the post light-support pipe and the electrical system ground. So a person using a garden hose near that post light might become part of the ground path and receive a shock.

Fault in appliances

To remedy an electrical fault in a building, first determine whether it is an equipment fault or a wiring fault. An equipment fault is an electrical failure in a piece of the electrical equipment, such as an appliance. A system fault is an electrical failure within the building wiring system. Since more faults occur in appliances than in wiring, check first for an appliance fault. Disconnect all appliances to see whether that eliminates the fault. It is easy to disconnect appliances since most of them are of the plug-in type. If the circuit remains normal and circuit breakers do not trip or fuses do not blow, the fault is obviously in one of the appliances. Then examine the appliances one-by-one to locate the fault.

Faults in wiring systems

If the fault is in the wiring system, the breaker will trip when the circuit is tested with the appliances disconnected. The next logical step is to open the outlets at lights, at switches, and at receptacles to try to locate the trouble. If nothing can be found, break the circuit by opening one of the splices, then retest. If the circuit holds, the fault is in the part of the system which was disconnected. To keep from tripping the circuit breaker or blowing the fuse under fault conditions some resistance can be added to the circuit as shown in Fig. 10-26. Adding such a test light does not remove the fault but it does add enough resistance to the circuit to keep the breaker from tripping. When there is a fault, the lamp will light up. Assume that there is a fault at one of the outlets fed from junction box A, for instance. When the circuit is opened at junction box B, the test light will remain lit, indicating that the fault is on the line side (circuit-breaker side) of line B. When the circuit is opened at junction box A, the test light will go out, which will indicate that the fault is on the load side of A. With this troubleshooting method, you have narrowed down the fault to the cables feeding from A and can now presume that the cables feeding from B are good.

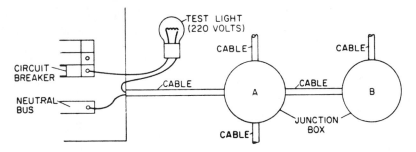

Figure 10-26   Use of test light in troubleshooting

If the system uses plug fuses, screw in the test lamp in place of the blown fuse. Remember, this test is not valid if there is any load on the circuit. Disconnect all lamps, appliances, and motors first. Also, as a safety measure always shut off the power while disconnecting circuits and be sure that the ungrounded wires are not shorted to ground or a metal outlet box.

Ground fault detection

A ground fault may be detected easily by using a clamp-on meter which can usually measure ac voltage, ac current, and resistance. Test leads are used to make ac voltage and resistance measurements. When current is being measured, the jaws of the clamp-on meter are opened and placed around the wire. Hold the clamp-on meter so the jaws will close by themselves to insure an accurate circuit measurement. The circuit does not have to be opened to measure current as it does when a regular ammeter is used.

In Fig. 10-27(a) we study a 240 V single-phase motor that is assumed to have a ground fault between the motor winding and the motor frame (case). Because the conduit grounds the motor frame to the panelboard, the ground current flows in the conduit. In other situations the ground path might be through the earth, water pipes, building steel, or a grounding wire.

The jaws of the clamp-on meter are placed around both current-carrying wires. If there is no ground fault, the meter will register at zero. The current would be zero if the current flowing to the motor is equal to the current returning from the motor. But if a ground fault occurs, part of the current returns to the panelboard through the conduit, and so the currents flowing in the wires are unequal. Accordingly, the clamp-on meter would indicate a current which is the difference between the currents flowing in the two wires.

A schematic diagram of that circuit in shown in Fig. 10-27(b). Let's assume that the motor usually draws 12 A but that a ground fault has occurred and 3 A now flows through the ground path. The current in line a has increased from 12 to 15 A while the current in line b remains at 12 A. The difference between the currents is 3 A, which is the same as the current in the ground path. If the jaws of a clamp-on ammeter could be placed around the two lines, it would indicate 3 A. In this example the ground fault occurred near line a end of the motor winding. If the ground fault occurred exactly at the midpoint of the winding, the currents would remain balanced and both currents would increase. For example, both currents might increase to 14 A. The resistance of the ground path would determine how much current flows in the ground path.

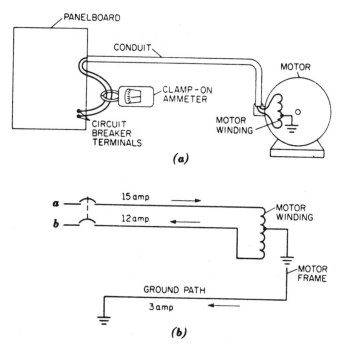

Figure 10-27   Method of detecting ground fault in motor: (a) layout diagram, and (b) schematic diagram

When this test is made, enclose all current-carrying wires inside the jaws of the clamp-on ammeter. If the circuit includes a neutral wire, remember to enclose it too because it carries current. Do not, however, include grounding wires, because they do not normally carry current.

If the circuit has a grounding wire, measure the unbalanced current with the grounding wire connected and then with it disconnected. There may be a change in unbalanced current due to the difference in resistance of the ground paths. When nonmetallic sheathed cable is used, the ground should disappear when the grounding wire is disconnected. In circuits that are either made with armored cable or are placed in conduit, there should not be any change in current.

10.17   *Finishing the Installation and Final Inspection*

When the rough wiring is installed, more wire than necessary is left in the outlet boxes. Those wires should be long enough to make connections easily, but not so long as to crowd the outlet box. Do not try to cut off the last little bit when the device is replaced; the new one may require slightly longer wires because of different terminal placement. Remember to strip off only enough insulation to make the connection. There must be no bare wire between the end of the insulation and the terminal screw.

After the device is wired, it is fastened to the device box by machine screws which pass through elongated holes in the strap of the device into the threaded holes in the ears of the device box. The enlarged ends, called plaster ears, may be cut off easily if they are not needed. Because device boxes usually are recessed slightly in the walls, the plaster ears lie on the wall surface and automatically hold the device flush with the wall surface.

Elongated holes are provided in device straps to permit devices to be mounted straight up and down when the device box is at an angle.

Faceplates are furnished with screws to fasten them to the devices. When plastic faceplates are used, tighten the screw very carefully to avoid cracking the plate. When the faceplates are used with two- or three-gang boxes, the devices have to be carefully lined up so that all screws will mate with their respective holes in the device straps.

Lighting fixtures

The lamp holders shown in Figs. 10-11 and 10-13 are mounted directly on the outlet box by machine screws which turn into the threaded ears of the outlet box. If the lighting fixture is too large to fasten directly to the outlet box, fasten a flat strap to the box instead. You may pass machine screws through holes in the lighting fixture and enter threaded holes in the strap. Other fixtures are made with one large hole which is of the correct size for passing over 1/8″ running-thread pipe, or nipple. The nipple may be screwed into the center hole of the strap shown in Fig. 10-28. Machine screws pass through the elongated holes in the strap to fasten the strap to the outlet box. The fixture is secured to the nipple by a decorative nut (finial). Another method of fastening such a fixture is to bolt the fixture stud as in Fig. 10-28(b) to the bottom of the box. An adapter is screwed into the stud. See Fig. 10-28(c). The small hole in the adapter is threaded to accept a nipple. The nipple-type assembly may be used with either ceiling or wall fixtures.

Fixtures which are larger but weigh less than 50 pounds may be fastened as shown in Fig. 10-29. The outlet box is supplied by two branch-circuit wires which are part of a two-wire cable secured by a cable clamp. The fixture stem screws into the hickey which is fastened by a locknut. Another

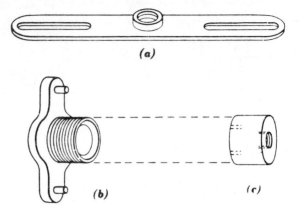

Figure 10-28   Lighting fixture mounting parts: (a) fixture strap, (b) fixture stud, and (c) adapter

type of hickey screws onto the fixture stud shown in Fig. 10-28(b). The fixture wires are spliced to the branch circuit wires. The canopy is then placed against the ceiling by a setscrew or some other device. Any combustible ceiling finish between the edge of the canopy and the outlet box must be covered with a noncombustible material, such as a piece of sheet steel.

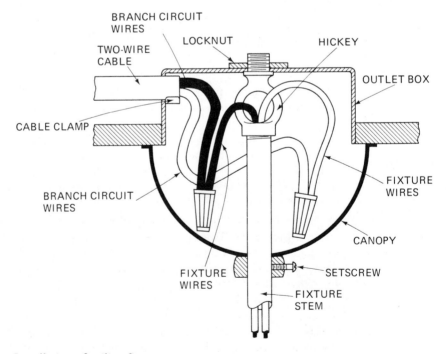

Figure 10-29   Installation of ceiling fixture

A fixture weighing more than 50 pounds must be supported independently of the outlet box. In new work, you may use bar hangers which include a fixture stud.

Completing the panelboard wirings

After all the outlets on a circuit are wired, the branch-circuit wires may be connected to the panelboard. Panelboards are not supplied with cable clamps, so it is necessary to use a bushing on each cable. Cables are often wired through knockouts on the side of the box. Select a knockout near the circuit breaker or fuse holder for the particular circuit to reduce the amount of wire within the panelboard. You must leave a longer piece of cable at the panelboard then at the outlet box because all neutral wires must be connected to the neutral bus. When you install nonmetallic sheathed cable, you also connect the grounding wire to the neutral bus. However, there may not be enough neutral terminals to connect each grounding wire individually. So splice the grounding wires together and connect only one wire to the neutral bus. When you install armored cable, remove the paint where the bushing will enter. Also, install the nut on the bushing so that makes it dig into the metal and so ensures a low-resistance grounding connection.

Plug-in circuit breakers are normally used for residential installations. Install a single-pole circuit breaker of the correct rating into the panelboard for each 120 V branch circuit. For 240 V branch circuits plug in two-pole circuit breakers. A single-pole breaker has one terminal for connecting the ungrounded wire of the branch circuit. A two-pole breaker has two terminals, one for each ungrounded wire. After all the circuits have been tested, lock the circuit breakers into position by the hardware provided.

When a fused panelboard has been installed, treat the neutral and the grounding wires in the same way as those of circuit breakers. There is a terminal near each fuseholder for plug fuses to connect the ungrounded wires of a branch circuit. Before installing the plug fuses insert the correct adapters, because type S fuses must be installed. The 240 V circuits are often protected by cartridge fuses on pull-out blocks. Wire the ungrounded wires as shown in the panelboard wiring diagram. You may have to run some short wires from the output side of the main fuses to the input side of the branch-circuit fuses to feed the 240 V branch circuits. If the pull-out for the main fuses has been plugged in, be sure to remove it before installing these wires. Make sure that the branch-circuit current will flow through the fuses, and that the branch-circuit wires and the wires from the main fuses are not connected to the same pair of terminals. Two cartridge fuses of the correct rating may then be placed in each pull-out block.

Circuit index

When all the circuits have been completed, and the appliances and heating system are connected and tested, and all breakers are installed (both active breakers and spares), you can install the panel trim. You should now mark up the panel index clearly and legibly. The index contains space for you to label each circuit. This task may seem unimportant, but its real value becomes apparent when the system develops a fault that has to be located by someone unfamiliar with the system.

Final inspection

At this point you should obtain the final inspection certificate. Usually the inspection is made automatically by an inspector sometime after the rough-wiring inspection is approved. Remember, it is important to obtain that certificate because it assures all the persons concerned that the wiring has been installed in a safe and secure manner. That protects the owner because it attests to the fact that the building is electrically safe and insurable without unusual risk. As a disinterested but qualified third party it also protects you the electrician when you can tell the owner that you have wired a system that is electrically safe for his house.

EXERCISES

10-1/   Define the terms "new work" and "old work" as they are understood in the electrical trade.

10-2/   What is meant in the electrical trade by "roughing-in"?

10-3/   How is a grounding clip installed?

10-4/   What is the color and shape of the grounding terminal on a receptacle?

10-5/   When both wires of a lighting fixture have the same basic color but one has a tracer, which wire is neutral?

10-6/   At what heights would you install receptacles?

10-7/   What factors should be considered in installing outdoor receptacles?

10-8/   What factors should be kept in mind when selecting switch positions?

10-9/   What type of lighting arrangement is recommended near a bathroom mirror? Why?

10-10/   Is the use of pendant-type lamps permissible in closets?

10-11/   What is the range of operating voltages for door chimes? For buzzers and doorbells?

10-12/   In a noncombustible wall, how must the front edge of an outlet box be positioned with respect to the final front surface of the wall?

10-13/   Describe two typical arrangements for mounting outlet boxes.

10-14/   Describe how armored cable is installed in an outlet box.

10-15/   Describe how nonmetallic cable is installed in an outlet box.

10-16/   Describe how the terminals of a standard duplex receptacle are identified.

10-17/   What is a two-circuit duplex receptacle? What arrangement is made on many standard duplex receptacles to permit easy conversion to the two-circuit type?

10-18/   In Fig. 10-9 the white wire is attached to a black wire. Under what circumstances, if any, is this permitted by the NEC?

10-19/   Describe a "three-wire" circuit. What are the advantages of this type of circuit?

10-20/   In a three-wire circuit suppose the two ungrounded wires carry currents of 18 and 22 A, respectively. What is the neutral current? Why?

10-21/   With what type of cable is a three-wire circuit wired?

10-22/   Draw the wiring-diagram symbols for a ceiling mounted pull-chain lamp holder. If three such lamps are used on a circuit, how is the last one indicated?

10-23/   Draw the wiring-layout symbol for a lighting fixture controlled by a single-pole switch. How would you indicate that the cable from the panelboard runs to the switch? To the lamp holder?

10-24/   Describe how you would install two lamp holders controlled by one switch.

10-25/   On a wiring diagram, how would you show the arrangement identified in Example 10.24?

10-26/   Describe how you would wire a pull-chain lamp holder fed from a switch that controls a switched lamp holder.

10-27/   Draw the wiring-diagram symbols for a switch with pilot light where (a) the cable feeds the switch, and (b) where the cable feeds the lamp holder.

10-28/   Describe two arrangements for installing a pair of three-way switches to control a single outlet.

10-29/   Describe the way you would connect a two-door chime.

10-30/   Define what is meant by fixed, portable, and stationary appliances.

10-31/   If a stationary appliance is rated at 400 W, does it require a special disconnecting means? Explain your answer.

10-32/   Does the NEC require a separate circuit for each automatically started motor, such as a water pump or oil burner?

10-33/   What are the voltage and current ratings of a receptacle commonly used for an electric clothes dryer? For an electric range?

10-34/   In using a standard two-wire cable to wire a water heater, what precaution must be taken to have the cable conform to NEC specifications?

10-35/   What size wire may be used for grounding a clothes dryer and how is the grounding accomplished?

10-36/   When a water heater branch circuit is protected by a plug fuse is any additional safety element required?

10-37/   What wire sizes are used to wire most electric ranges?

10-38/   Using a single 50 A cable describe how you would wire a wall-mounted oven and a counter-mounted cooking unit.

10-39/   When electric space heaters are supplied by a 30 A branch circuit, what is the current limitation? Why?

10-40/   Describe how a line-voltage thermostat operates to control an electric space heater.

10-41/   Describe how the overheat switch of an electric space heater is wired.

10-42/   Describe the thermal relay and how it operates in a low-voltage control system for electric space heating.

10-43/   Describe one method of checking individual branch circuits for a possible short-circuit.

10-44/   How is a continuity test made after the rough-in is completed?

10-45/   What instrument is best suited to testing insulation? Why?

10-46/   How is it possible for a ground fault to exist without blowing a fuse or tripping a circuit breaker?

10-47/   What step would you take first to determine where an electric fault might exist in a wiring system? Why?

10-48/   What is the easiest way to locate a ground fault?

10-49/   How would you install a fixture weighing more than 50 pounds?

10-50/   What types of plug-in circuit breakers would you use to complete the panelboard installation in a residence?

11
Old Work

11.1   *General*

As you know, "old work" is the wiring of buildings which were completed before the wiring is started. Such work presents few electrical problems which have not already been discussed in connection with new work. Most of the difficulties will be problems of carpentry or mechanics—how to run wires from one point to another with the least effort and with minimum damage to walls, floors, and ceilings. Old work requires more cable than new work because it is often better and easier to run the cable the long way around through spaces that can be easily used.

In old work, every building presents a special set of problems because of construction variations. At times it may seem that every carpenter has his own construction methods. If you study buildings that are presently being constructed, you will better understand what is hidden behind the plaster of older buildings. For example, the differences are especially noticeable in the construction around doors, windows, and corners and in the spacing between the different members of the structure.

11.2   *Making Wall and Ceiling Openings*

Cutting good openings for outlet boxes requires a certain skill and lots of common sense. The openings must not be too large and they must be made neatly. After the building structure has been studied and the wiring has been planned, mark the approximate place where the outlet box will be located. If possible, allow yourself some leeway so that the opening may be shifted slightly in any direction. To determine whether there is a stud or a joist in the way, sound or thump on the wall and ceiling. Thumping will produce a different sound at a timber than in the hollow between timbers. In older housing the plaster is supported by narrow wooden strips called laths. The space between the laths is filled with plaster. A newer house may have been plastered over metal lath or dry wall, such as gypsum board, hence, the difference in sounds.

When the house is plastered with wooden lath, dig through the plaster at the approximate outlet location until you determine the space between two laths. Then bore a small hole all the way through the probe and a stiff wire to determine whether a timber is nearby. If the space is clear, hold the outlet box against the wall and make a mark around it. If you make a proper layout for a device box only one lath will be completely cut off, as shown in Fig. 11-1(a). A side view of the installation is seen in Fig. 11-1(b). You must locate the center of the box on the center of a lath, not in the space between laths. Drill holes carefully at diagonally opposite corners large enough for a hacksaw blade to enter. Use the hacksaw blade with the teeth pointing toward you—the opposite of the usual method. You cut when you pull the hacksaw blade toward you. If

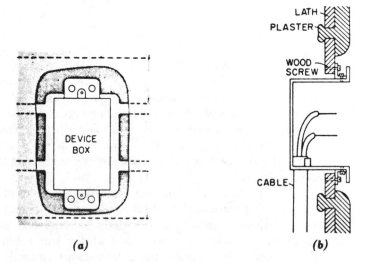

(a)

(b)

Figure 11-1   Proper location of device box with respect to wood laths: (a) front and (b) side views

you push the blade in the usual way, there is a danger that the lath will pull away from the plaster and leave some plaster unsupported. While you saw, hold your hand against the plaster to support it and to reduce the tendency for the plaster to pull off the wall.

Where dry-wall construction has been used, a portable jigsaw, or a saber saw, may be used to make the openings. Equip the saber saw with a metal-cutting blade and use a low cutting speed.

Be careful not to mar the walls or get them dirty. If possible, confine the cutting to the areas which will later be covered by the lighting fixture canopy or the wall plate. If the walls are papered, a repairable hole can be made by first cutting a cross in the paper with a sharp razor blade and then folding back the triangular pieces thus formed. If water will not damage the wallpaper, use water to loosen the sections to be folded back.

11.3   *Floor Openings*

Sometimes outlet and switch boxes may be located so near obstructions that it is necessary to lift the boards of the floor above. Attic flooring made from plain boards is easily lifted because the location of the floor joists can be determined from the nailheads on the surface. Other floors usually are made from tongue-and-groove boards with concealed nailing, as shown in Fig. 11-2(a). To lift these boards, it is necessary to cut off the tongue of the board with a very thin chisel, such as a putty knife with the blade cut off short and then sharpened. Drive this chisel down between the two boards to cut off the tongue. The floor joists will be located where the chisel will not penetrate more than an inch. Cut off the tongue for the entire distance between three joists, because the floor will then be stronger when the flooring is replaced. Drive small holes at two points next to the joists. You may use a carpenter's square and a pencil to draw lines at right angles to the flooring and parallel to the joists. Cut off three boards with a saber saw or a keyhole saw beginning at the small holes. A keyhole saw has a small blade which tapers to a point. The electrically-operated saber saw will cut faster, but it tends to splinter the surface of flooring. After the tongue is removed insert a chisel at a number of points along the board in order to pry it up. Pound on the adjacent board with a hammer to help remove the first board. Use a block of wood under your hammer to protect the floor from being damaged.

After you remove the first board, the other boards may be easily lifted to provide access for wiring. Before you replace the floor boards, nail cleats on the sides of the joists to support the ends of the floor boards as shown in Fig. 11-2(b).

Remove the nails from the floor boards and joists before replacing the floor boards. Replace each piece in its original position and nail it to the joists and cleats with finish nails. Finish nails have a small head about twice the diameter of the main part of the nail, whereas common nails have a larger flat head. Especially in hardwood floor boards, the nail will be easier to drive if you

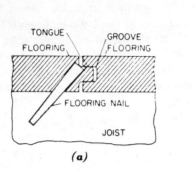
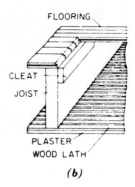

(a)   **(b)**

Figure 11-2   Typical floor construction: (a) tongue-and-groove flooring, and (b) cleat to support cut end of flooring

first drill nail holes through the floor boards. Use a drill about the same diameter as the main stem of the nail. After the head of the finish nail is flush with the flooring, you may countersink it by using a carpenter's nail set and a hammer. You may blend in the nail holes and saw cuts with the floor boards if you fill them with one of the soft, colored, putty-like compounds which are made in many shades and sold at most hardware stores and lumber yards.

11.4   *Mounting Outlet Boxes*

For new work all outlet boxes must be at least 1-1/2″ deep; for old work shallower boxes may be used if a deeper box would seriously weaken a member of the building. For example, recessing a 1-1/2″ deep box into a floor joist would seriously weaken that joist. The outlet box should be large enough to accept the wires without their being jammed together. Never use outlet boxes less than 1/2″ deep. A typical shallow box for use with nonmetallic sheathed cable is shown in Fig. 11-3. Shallow boxes are also available with a built-in fixture stud.

Although a shallow outlet box may be fastened directly to wood lath with screws, it is better to fasten it to a joist. If the flooring is removed above the place where you want to mount the outlet box you can nail a wooden block with a nominal thickness of at least one inch between the joists to support the lighting fixture. Outlet boxes may be supported between joists by using the old-work hanger shown in Fig. 11-3(b). Slip the hanger through a hole in the ceiling large enough to clear the bar and the stud. Then pull the wire until the stud is centered in the hole. The bar should be at right angles to the lath for the best support. The stud passes through a hole in the bottom of the outlet box where you have removed a knockout. Use a locknut to secure the outlet box to the stud.

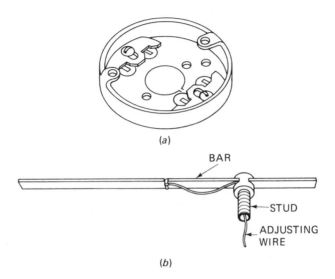

(a)

(b)

Figure 11-3   (a) Shallow outlet box and (b) old work hanger

11.5   *Mounting Device Boxes*

The device box shown in Fig. 11-1 may be fastened to the wood lath by four small wood screws passing through the holes in the mounting ears. The mounting ears, which are adjustable so the front edge of the outlet box may be made flush with the wall, are shown more clearly in Fig. 11-4. When the device box is mounted on wallboard or similar materials, you can not use screws because the material is too weak to hold them. Install the device box instead with the device-box supports shown in Fig. 11-4(a). These supports consist of a pair of metal straps with ears. Note that the projection of the strap beyond one ear is longer than that beyond the other. Slip the device box into the opening until it is restrained by its mounting ears. Next, insert the long leg of a strap until the ear rests against the edge of the opening. Then it is possible to insert the short end. Move the strap until both ends are hooked behind the wall board; pull the strap up firmly by one ear; bend the ear down inside the box. Repeat the process with the other ear and strap. Be sure the ears remain close to the inside surface of the box or grounds may result.

The device box in Fig. 11-4(b) is easy to install, but it requires a larger hole to clear the metal strips which are on the sides of the box. After you slip the device into the wall opening, tighten the screws on the sides since they cause the metal strips to expand outward and grip the wall. Be careful. Because the expanding metal strips are not directly opposite each other, the device box tends to twist in the wall opening as you tighten the screws.

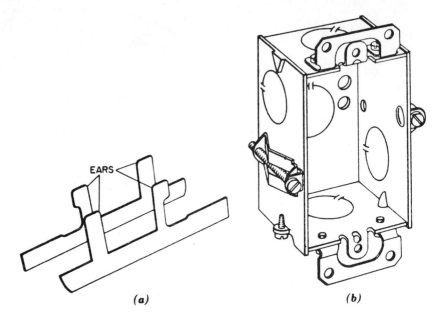

EARS

(a)   **(b)**

Figure 11-4   (a) Device-box supports and (b) screw-clamp device box

11.6   *Running Cable*

In old work, you install the cable after the outlet box openings are cut and before the outlet boxes are installed. Use nonmetallic sheathed cable and armored cable normally because they are flexible and easily fished into the empty spaces. In general, you will find that it is easier to run cables from the cellar or the attic into a wall than through other parts of the building.

Removing the baseboards in the room above will often make it possible to reach a ceiling outlet. Ceiling outlets generally do not line up with switch outlets, so removing the baseboards makes it possible to run a cable across the wall without disturbing the finished part of the wall. Cut a groove in the plaster where it will be covered by the baseboard. If two wall outlets in the same room are to be connected, remove the baseboard to help in running the cable.

The exterior walls of newer homes can present a problem not encountered in older homes. For instance, fishing cables is difficult in residences in which insulation has been installed to make heating and air conditioning more economical. Fiberglass insulation should be handled carefully; it can be irritating to the skin and dangerous to the eyes.

Fish tapes are useful in installing the cable because the fish tapes can often be pulled from one opening to the other where it would be difficult to push a cable. Another efficient method is to feed a fish tape in from each opening and hook the ends together. One fish tape is used to pull the other fish tape and the cable through the openings. A small weight on a string is also useful for fishing between the walls of partitions.

11.7   *Making Connections*

At the new outlet, the general procedure is to prepare the end of the cable as it would be prepared for new work and then secure the cable in the outlet box. Next, work the outlet box with the cable back into the opening and then fasten it to the wall. An oversize opening makes installation of the box easier, but it may also make it difficult to fasten the box properly. A smaller opening is a sign of good workmanship. The opening around the outlet box should later be filled with patching plaster or plaster of paris.

Make connections to the new device in the same way as for new work. Remember to include a grounding wire and install grounding receptacles where it is practical to obtain a good ground. When you must connect new wires to the old wires, the neutral and ungrounded wires may not be easy to identify. Be careful—because the neutral wire may appear black, it is best to determine which wire is actually the neutral.

One method of testing is shown in Fig. 11-5 where a neon circuit tester

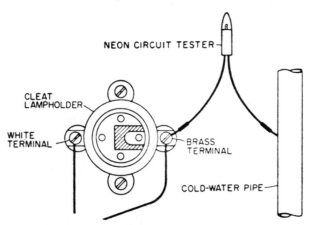

NEON CIRCUIT TESTER

CLEAT LAMPHOLDER

WHITE TERMINAL

BRASS TERMINAL

COLD-WATER PIPE

Figure 11-5   Use of neon circuit tester

is connected between the brass terminal on a cleat lamp holder and a cold-water pipe. The lamp in the circuit tester will be illuminated with a pink glow if there is normal voltage between the brass terminal and the cold-water pipe. Use caution when making this test so that you will not become part of the circuit and receive a shock. The neon circuit tester may be used if no cold-water pipe or ground is available nearby. Because a high resistance is included in the neon circuit tester, one lead can be grasped safely in your hand while the other one is touched to the terminals. The terminal at which the neon lamp glows is the ungrounded one. However, the neon bulb will not glow as brightly as when it is connected to a good ground.

Remove the fuse or turn the circuit-breaker off before you connect the new wires to the old ones. Make any splices or connections well and insulate them properly.

When you add outlets to older homes, you may find a type of wiring which has not been discussed because it is not ordinarily used in new installations. This wiring is called knob-and-tube work and uses individual wires rather than cables. When these old wires pass through a timber, such as a joist, they are protected by porcelain tubes; when they run along a surface, they are supported by cylindrical porcelain insulators called knobs. Consult the NEC which lists the requirements for knob-and-tube wiring.

11.8   Surface Wiring

Surface wiring is another method of adding outlets to old wiring. In this method you use surface-mounted outlets and cables. Use is made of a special plastic two-wire cable which is fastened to the walls by driving small round-head nails through holes which have been prepunched in the cable. The cable is flexible enough that it can be bent in either direction to go around corners. Surface-mounting switches, lamp holders, receptacles, and junction boxes are all made for use with the cable. A pull-chain lamp holder used with this cable is shown in Fig. 11-6.

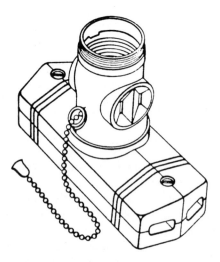

Figure 11-6   Surface-mounting pull-chain lamp holder

11.9 *Modernizing the Installation*

Many residential wiring systems are obsolete because they do not have enough circuits to supply all the appliances presently in common use. The general purpose, or lighting, circuits usually remain adequate for the intended loads. It is the additional kitchen and laundry appliances and air conditioners that are most likely to overload the circuits. Some appliances even operate from 240 V circuits which may not be available in a given building. The service-entrance equipment is also likely to be too small for today's load.

Before deciding that all the old wiring needs replacement, analyze the load. If all the appliances were disconnected, the general-purpose circuits for lighting and receptacles for loads such as radios, television sets, and vacuum cleaners are probably adequate. It is easier and cheaper to modernize the wiring system if the existing branch circuits can be used. The work is likely to require adding a new service entrance, kitchen circuits, laundry circuits, and other appliance circuits.

Plan the installation to determine how many additional branch circuits are needed and the size of service-entrance equipment. Allow two 20 A branch circuits for the kitchen receptacles. Also, plan for individual branch circuits for other appliances—like furnace motors, air conditioners, and ranges. The service-entrance equipment should be large enough so that additional circuits may be added later.

11.10 *Service-Entrance Changes*

Check with the power supplier for the recommended procedure for changing over to a new service entrance. Knowing the power company practices will help you to improve the wiring with the least inconvenience to your customer. The general practice is to install the new branch circuits without connecting them. The new service-entrance equipment with a current rating of at least 100 A can be installed near the present service-entrance equipment unless a better location is available. You will have to have the new service entrance inspected before the power company will connect the service-drop to it.

One method of connecting the new panelboard to old wiring is to completely remove the old panelboard and replace it with a junction box as shown in Fig. 11-7(a). Because the old wires or cables will usually not be long enough, it is necessary to lengthen them by splicing cables to them. The splices are made in the junction box. All circuit breakers or fuses are in the new panelboard.

Another method is to use the old panelboard as a distribution center. See Fig. 11-7(b). A three-wire cable protected by a two-pole circuit breaker connects the two panelboards. If the old panelboard contained main fuses, or service overcurrent devices, you may leave them in position. Connect the new three-wire cable below the main fuses but ahead of the branch circuit fuses.

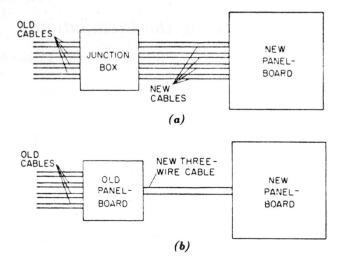

Figure 11-7   Basic connection methods in rewiring installations: (a) junction box replacing old panelboard and (b) using old panelboard as a distribution center

Make sure that the three-wire cable is heavy enough to carry the load current. If the existing service-entrance equipment is rated at 30 A, then a three-wire No. 10 cable protected by a two-pole 30 A circuit breaker is adequate. If the existing service-entrance equipment is rated at 60 A, then a three-wire No. 6 cable protected by a two-pole 50 A circuit breaker is adequate. An alternate method is to replace the old panelboard with a new distribution center which contains circuit breakers with the proper ratings.

EXERCISES

11-1/   In a house plastered with wooden lath, describe how you would go about making a wall opening.

11-2/   In cutting through a wall plastered with wooden lath, how is the hacksaw blade held? Why?

11-3/   With respect to the laths, how should a box be located in a plastered wall? Why?

11-4/   How can a repairable hole be made in wall paper?

11-5/   Describe how you would lift floor boards when the floor is made from tongue-and-groove boards.

11-6/   What should you do before attempting to drive finish nails in a hardwood floor?

11-7/   What should you do after driving finish nails in a hardwood floor?

11-8/   What is the minimum depth of an outlet box for new work? For old work?

11-9/   Describe how a device box should be mounted in old work.

11-10/   What type of cable is normally used in old work? Why?

11-11/   When insulation has been blown into the walls of a newer home, what precautions should be taken? Why?

11-12/   Describe a test for determining which wire is neutral.

11-13/   How is it possible to use a neon circuit tester when no suitable ground is available?

11-14/   Describe tube-and-knob work.

11-15/   What is "surface wiring" and what type of components are required for work of this type?

11-16/   In modernizing a residential electrical system, where are modifications usually required?

11-17/   If a modification of an old service entrance is needed, what is the minimum capacity the NEC permits for the new service entrance?

11-18/   Describe two methods for connecting a new panelboard to old wiring.

11-19/   In doing old work, what allowance should be made for the kitchen?

11-20/   If, in connecting a new panelboard to old wiring, you find that splices are needed, where must they be made?

Part
5

12
Motors

12.1   *General*

An electric motor is a device that converts electrical energy into mechanical energy. All electric motors operate on the same basic principle—the interaction of two magnetic fields to produce mechanical rotation.

12.2   *Basic Motor Action*

Figure 12-1 illustrates the action of a current-carrying conductor in a magnetic field. Assuming that the current through the conductor is directed into this page, as you look at it, the magnetic field surrounding the conductor has its lines of force running counterclockwise. The magnetic field provided by the poles of the permanent magnet are, of course, directed from north to south.

Lines of force resemble elastic bands in that they always try to contract to their shortest length. In Fig. 12-1, the lines of force of both fields are in the same direction *under* the conductor, so they add. Above the conductor, however, the lines of force are in opposition so the total field in that vicinity is weakened. The crowded lines of force below the conductor try to straighten out and, in so doing, exert an *upward* push on the conductor.

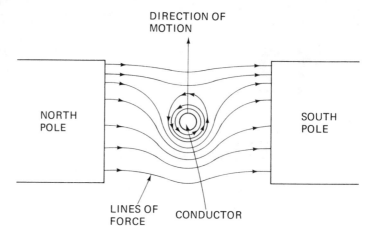

Figure 12-1   Action of a current-carrying conductor in a magnetic field

Now let us apply this principle to a dc motor to see how rotation of a motor shaft is produced. A simple dc motor is illustrated in Fig. 12-2. Here, a single loop is located between the poles of a permanent magnet. The ends of the loop connect to metallic commutator *segments* which are separated by a small air space. Connection to the external circuit is made through carbon *brushes* which seat on the commutator segments. Current flows from the negative terminal of the dc supply, through commutator segment and brush A, the loop, brush B and its associated commutator segment, and then back to the positive terminal of the supply.

Refer now to Fig. 12-3. The + mark in the center of the conductor represents the tail of an imagined arrow and shows current *flow into* this book page. The dot in the center of the opposite conductor represents the arrow's head and shows current *flow out* of our "demonstration" page. With the sides of the loop positioned as shown in Fig. 12-3(a), there is an upward thrust on one side of the loop and a downward thrust on the other side. The resulting twisting force exerted on the loop is called *torque*.

When the sides of the loop are perpendicular to the lines of force created by the permanent magnet, the two segments of the commutator are short-circuited by the brushes. As a result, essentially no current flows through the loop. The point at which this occurs is called the *point of commutation*. Seemingly, at this point, our single-loop generator would come to rest. The inertia of the loop is great enough, however, to carry it past the point of commutation and the situation depicted in Fig. 12-3(c) then exists. Although the sides of the loop have reversed position, the direction of current through the loop has not. Thus, the loop continues to rotate in a clockwise direction.

In an actual working motor, unlike the theoretical motor we have just examined, the single loop is replaced by a group of *multiple loop* coils, called *armature* coils, and each armature coil is connected to a set of commutator

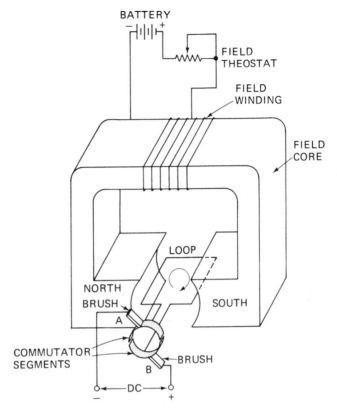

BATTERY

FIELD THEOSTAT

FIELD WINDING

FIELD CORE

LOOP

NORTH

BRUSH

SOUTH

A

COMMUTATOR SEGMENTS

BRUSH

B

DC

Figure 12-2   A simple dc motor

segments which are electrically insulated from one another by strips of mica. The coils are mounted in slots in an iron core called the *armature*. In place of the permanent magnet, all but very small motors employ *electromagnets* to provide the magnetic field in which the current carrying armature loops rotate. A metal shaft, which passes through the center of the armature, is supported at both ends of the motor housing in such a manner as to permit easy rotation.

The torque that would be produced by a single-loop motor would, of course, be very small. Moreover, that torque falls to zero twice during each rotation of the loop; that is, when the loop is passing through the point of commutation. Not only does the practical motor produce the torque needed to do a meaningful amount of work, but the armature coils are actually arranged in such a way that torque never drops to zero.

Some large motors use multiple electromagnets to increase the torque. These electromagnets are called *poles* and there is always an *even* number of poles. The poles are attached to the motor housing and, because they do not rotate, the poles and housing are called the motor *stator*. The armature, accordingly, is called the *rotor*. In all multi-pole motors, the armature coils are

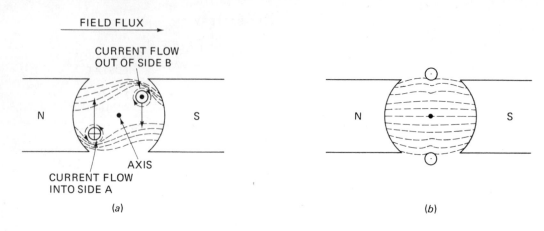

FIELD FLUX

CURRENT FLOW
OUT OF SIDE B

N   S

AXIS
CURRENT FLOW
INTO SIDE A

(a)

N   S

(b)

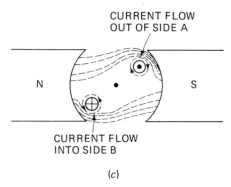

CURRENT FLOW
OUT OF SIDE A

N   S

CURRENT FLOW
INTO SIDE B

(c)

Figure 12-3   Loop rotation as a result of field interaction

spaced so that opposite sides of a given armature coil pass beneath opposite-polarity poles at all times.

Direction of rotation

When the direction of the current flow through either the armature or field of a dc motor is reversed, the direction of motor rotation is also reversed.

The reversal of armature current changes the direction of the lines of force around the current-carrying conductors. Careful inspection of Fig. 12-3 indicates that, when this occurs, the direction of rotation must reverse.

When current through the field windings is reversed, the poles change polarity, and the resulting interaction between the field and armature again causes a reversal in the direction of rotation.

Counter EMF

When a conductor cuts lines of force, a voltage is generated in the conductor. You will recall from Chapter 1, Sec. 1.20 of this text that because this voltage opposes the voltage applied to the motor, it is called *counter electromotive force,* or CEMF.

The magnitude (strength) of the CEMF depends on the *number* of armature conductors, the *speed* of rotation, and the *strength* of the magnetic field. In any given motor, of course, the number of conductors is fixed, but the speed of the motor and the strength of the magnetic field may be varied at any time. Thus, the magnitude of the CEMF is controllable. Either an increase in speed or an increase in the number of magnetic lines provided by the field act to increase the CEMF.

The amount of CEMF generated by a motor regulates the armature current. Because the resistance of an armature is very low, CEMF is the main factor limiting the flow of armature current in a dc motor. When the armature is turning slowly and generating only a small CEMF, there is a large armature current which causes the motor to have strong torque. This makes the motor speed up as it overcomes the opposition of the mechanical load. As the motor speeds up, it generates more CEMF. This then decreases the armature current, decreases torque, and decreases the speed. The motor will turn at a speed that will produce just enough torque to balance the mechanical load on the motor.

Neutral plane

A plane perpendicular to the lines of force of the field is called the neutral plane. For proper commutation, each armature coil must be in the neutral plane when it is shorted by the brushes. This seemingly simple condition is not quite so simple to achieve in practice.

Armature reaction

When current passes through the armature, the interaction between the lines of force of the armature and the field is such as to distort the field. This distortion causes a shifting of the neutral plane and is called armature reaction. Since current through the armature varies with the load, the neutral plane also varies with the load. If each coil of the armature is not in the neutral plane when it is shorted by the brushes, excessive sparking occurs at the brushes and this sparking causes pitting of both the brushes and the commutator segments.

It is not practical, of course, to shift the position of the brushes each time a change in the position of the neutral plane occurs. It is practical, however, to use small interpoles (also called commutating poles) between the main poles. The windings of the interpoles are in series with the armature winding. With this arrangement, the strength of the interpole field changes with changes in armature current and cancels the distortion caused by armature reaction. Thus, the interpoles improve commutation under varying loads.

Although the interpoles correct for field distortion due to armature reaction, the field is weakened as a result of armature reaction. To keep the field strength constant under varying-load conditions, some motors have compensator windings which are connected in series with the armature and interpole windings and are placed in slots on the faces of the pole pieces.

Torque and horsepower

As you know, the twisting force produced by a motor is called torque. The amount of torque produced by a motor is proportional to the armature current and the strength of the field. Increasing either the armature current or field strength increases torque. Maximum torque is usually produced when the motor stalls as a result of excessive load because the rotor is standing still, no counter electromotive force is produced, and armature current is maximum.

While the amount of torque a motor produces is an indication of how powerful the motor may be, a torque rating *does not* tell us how much *work* a motor can do. To do work, a motor must move a load. A stalled motor may be producing a large torque, but no work gets done unless the motor turns. For this reason, motors are rated according to their capacity to do work—in horsepower.

Horsepower (hp) is the unit rating that expresses the mechanical power of the motor. For example, a motor is rated at 1 hp when it can lift the equivalent of 33,000 pounds one foot in one minute. It might also lift 16,500 pounds 2 feet in one minute, or 11,000 pounds three feet in one minute, or any other proportionate equivalent.

As indicated on the motor nameplate, a given motor will produce a given horsepower at a *rated* speed.

As noted in Chapter 1, Sec. 1.10, the electrical equivalent of mechanical horsepower is expressed in watts; 746 watts is equal to 1 horsepower. A motor with 746 watts electrical power input would not, however, deliver 1 hp to the load at the output shaft. Mechanical and electrical losses in the form of heat and friction use up a portion of the input power. The actual output will be the input power multiplied by the efficiency rating of the motor (a factor which depends on the design of the motor).

Speed regulation

The ability of a motor to maintain its rated speed under load is called its speed regulation and is expressed as a percentage. Mathematically,

$$\% \text{ regulation} = \frac{\text{no load speed} - \text{rated load speed}}{\text{rated load speed}} \times 100 \qquad (12\text{-}1)$$

A small percentage indicates good speed regulation. If, for example, a given motor runs at 1500 rpm with no load and slows to 1450 rpm with its rated load, its speed regulation is

$$\% \text{ regulation} = \frac{1500 - 1450}{1500} \times 100 = 3.33 \text{ percent}$$

12.3   *DC Motor Types*

There are three common types of dc motors. Each type has different torque and speed characteristics.

Series motor

As shown in Fig. 12-4, a motor of this type has its field and armature connected in series. Thus, any current through the armature must also flow through the field winding. The field winding consists of a few turns of heavy wire of low resistance.

The series dc motor is noted for its high starting torque which it develops as a result of the heavy current through both the field and armature windings when the motor is stalled. When the motor starts, current is limited only by the low resistance of the armature and field windings. Since torque is proportional to armature current and field strength, a heavy armature current increases both factors that make up torque.

Whenever the load on a series motor is increased, the motor slows down and develops more torque. The reason it slows down so much is that the increased load increases the armature current, which also flows through the field winding and strengthens the field. With a strong field, the armature develops more CEMF. The increased CEMF then limits the armature current. If regulated to turn more slowly, the armature can draw enough current to develop the torque needed to balance the mechanical load.

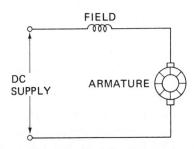

Figure 12-4   Schematic diagram of a series motor

The series motor has very poor speed regulation and will run away if it is operated without a load. Under no-load conditions, the motor develops a high starting torque and picks up speed rapidly. As the armature rotates, CEMF limits armature current, but this also weakens the field strength. Now the armature must turn even faster to generate the same CEMF, but each increase in speed further weakens the field. The process is cumulative and will lead to destruction of the motor. For this reason, series dc motors are coupled directly to their loads.

Series motors are used in heavy industrial applications and are not encountered in residential wiring systems.

Shunt motors

As shown in Fig. 12-5(a), the field winding of a *shunt* motor is connected in parallel with the armature. Thus, the full supply voltage is applied to both windings. The field consists of many turns of wire because the supply voltage is applied to it. The resistance of the armature is low and current through it is limited mainly by CEMF.

The shunt motor has nearly constant speed from no load to full load, and is often called a *constant-speed* motor. Under no-load conditions the motor speeds up enough to produce sufficient CEMF to limit armature current to a safe value. Because the field coil and armature are in parallel, the field strength is almost independent of armature current. At a particular no-load speed, the CEMF almost equals the applied voltage and there is no further increase in speed. As load is applied, the motor slows down slightly and generates less CEMF. This allows more armature current to flow and the motor produces more torque to balance the increased load. Since a small change in CEMF produces a large change in armature current, the motor does not slow down appreciably.

Another factor which helps the shunt motor maintain its speed under load is the weakening of the field as a result of armature reaction. As armature current increases, armature reaction increases, and the field becomes weaker. A weaker field causes less CEMF for a given speed and the motor does not have to slow down very much to produce the necessary torque.

A rheostat connected in series with the field coil can control the speed of a shunt motor. Any increase in resistance causes a decrease in current through the field winding and makes the field weaker. With a weaker field the armature generates less CEMF. This allows more current to flow, and that, in turn, tends to produce more torque. The increased torque causes the armature to speed up until the torque produced by the motor balances the mechanical load.

When the resistance is decreased, more field current flows and the field becomes stronger. Now the CEMF is increased and less armature current flows. With less armature current, the motor does not produce enough torque to balance the load, and the motor slows down. Now the CEMF decreases, allowing enough current to flow to produce the necessary torque to balance the load at reduced speed. We can sum up these two conditions by saying that decreasing the field strength causes the motor to speed up, and increasing the field strength causes the motor to slow down.

The field winding of a shunt motor should never be opened while the motor is running, or the motor may run away. As you know, the motor speeds up as the field is made weaker. If the field winding is opened, there may be enough residual magnetism in the pole pieces to give the motor a very weak field. The motor will accelerate and may reach a speed that will cause its destruction.

The shunt motor has good starting torque, though not as good as a series motor. When the motor is started, armature current is limited only by the resistance of the armature. Thus, a large armature current flows and produces a large torque. Because the field strength is almost constant, the torque produced is almost directly proportional to armature current.

The speed and torque characteristics of a shunt motor are shown in the graph in Fig. 12-5(b). The speed curve shows that the speed remains almost constant over a wide range of current. As current increases with an increasing load, the speed falls off slightly. Notice that the torque curve rises steadily as the current increases.

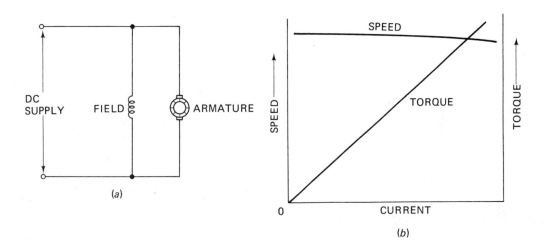

Figure 12-5   Shunt motor: (a) schematic diagram and (b) speed-torque characteristics

The shunt motor is used where constant speed under varying load is important. Also, since it is easy to adjust the speed by changing the field strength, the shunt motor is used where motor speed must be adjusted accurately.

Compound motors

The compound motor has both a series-field coil and a shunt-field coil. These two fields give compound motors characteristics that are a combination of the series- and shunt-type motors. The schematic diagram of a compound motor is shown in Fig. 12-6(a). Notice that a rheostat is used to adjust the strength of the shunt field. Because the series field carries the total current of both the armature and shunt field, it is made up of only a few turns of heavy wire. If the series field is connected so that its magnetic field aids that of the shunt winding, the motor is called a cumulative compound motor. If the series field is connected so that it opposes the shunt field, the motor is called a differential-compound motor.

The cumulative-compound motor resembles the series motor because it has a high starting torque, and it also resembles a shunt motor because it has a maximum speed. When it is loaded, the series field increases the total field strength and causes the torque to increase. At the same time, the stronger field causes the motor to slow down. Under no-load conditions, the presence of the shunt field ensures that the motor will have a maximum speed that it will not exceed, just as in the shunt motor. The cumulative-compound motor is used where high starting torque is needed, but the runaway characteristic of a series motor cannot be tolerated.

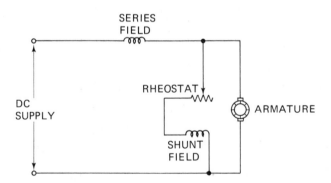

Figure 12-6   Schematic diagram of a compound motor

The differential-compound motor tends to speed up when a load is applied to it. The load causes increased current through the series field coil. Since the series field opposes the shunt field, the total field is weakened, and the motor tends to speed up. The differential-compound motor is seldom used because its speed is unstable. When starting under load it is possible for the series field to overcome the shunt field, and the motor can run away, or even reverse direction. For this reason, the series field coil is usually shorted-out while the motor is being started.

Of the various types of motors described, the only one you are likely to encounter in residential or farm wiring is the shunt motor.

12.4   *Motor Starters*

Small motors (1 hp or less) are usually started by placing the motor directly across the power-supply line. The motor draws several times the normal running current, but the resistance of the winding is sufficient to limit the current to a safe value until the motor gets up to speed. When the motor is running, CEMF opposes the applied voltage and limits the armature current to its normal value.

Motors larger than 1 hp generally employ a starting device which limits starting current to a safe value.

Most starting devices are simply variable resistances that can be inserted in series with the motor winding. The resistors limit the initial current to a safe value and are cut out of the motor circuit, one by one, until the motor is finally connected directly across the line. Starters may be designed for either manual or automatic operation.

A *starting box* intended for manual operation with a shunt-type dc motor is shown in Fig. 12-7. The diagram is drawn so as to show the mechanics of operation as well as the circuitry. One side of the supply is connected directly to the armature and the field coil through switch *S*. The other side of the supply is connected to the other side of the armature and the field coil through switch *S* and the starting box. The armature is connected to point 5 on the starting resistance *R*, and the field coil is connected through a so called *holding coil* to point 1 on resistance *R*. A *wiper arm, W,* is connected to switch *S*.

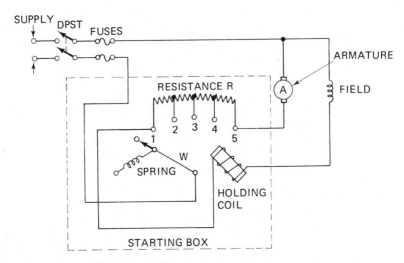

Figure 12-7   Schematic diagram of a starting box intended for manual operation with a shunt-type dc motor

When the wiper arm is in the position shown in Fig. 12-7, no current can flow either to the field or to the armature even with switch S closed. Thus, closing switch S has no immediate effect on motor operation. If we close switch S and then move the wiper arm to position 1, however, current will flow to the motor through two paths. Current will flow from position 1, through the holding coil to the field winding, then to the other side of the line. This places full line voltage on the field winding, allowing maximum field strength for starting.

The other current path is from position 1 through resistance R and the motor armature to the other side of the line. Thus, the resistance limits the armature current but has no effect on the field current. The armature current produces a torque, and the motor starts to turn. As it turns, the CEMF builds up, and reduces the armature current until the motor arrives at the speed determined by resistance R and the CEMF. At this point the motor speed cannot increase unless the armature current is increased.

Moving wiper W to position 2 removes the resistance between positions 1 and 2 from the armature circuit and allows more current to reach the armature. This increased armature current allows the motor to turn faster, but at the same time builds up a stronger CEMF which eventually limits the speed once again. By moving the wiper arm to position 3 we again get an increase in armature current, speed, and CEMF. Thus, as the wiper arm is moved from position 1 through position 5, both speed and CEMF are built up gradually until the normal operating speed is obtained.

The progressive movement of the wiper arm places the resistance of the starting box in series with the field coil and a decrease in field current occurs. Because the resistance of the starting box is very small compared to the resistance of the field winding, the decrease in field current is slight.

At position 5, none of the resistance is in the armature circuit, and the motor is running at full speed with a normal CEMF. At this point, the holding coil, which is in series with the field winding, acts to hold the wiper arm in position 5; hence its name. The holding coil may work in one of two ways: it may either hold the wiper arm directly by electromagnetic attraction between the core of the coil and the wiper arm itself, or the starting box has a mechanical latch that actually holds the wiper. In the latter case, the latch is held in position by the magnetism of the coil.

In either case, if the coil is suddenly de-energized, the wiper arm is released. When it is released, the spring tension on the arm returns it to the off position, removing all current from the motor. This gives us an automatic undervoltage protection and also ensures that if the field circuit opens, the armature circuit will also open so that the motor cannot run away. To stop the motor, we merely open switch S, which removes all current from the motor circuit. The wiper arm then moves to its off position, resetting the starting box.

In operating a starter of the type described, the wiper arm must not be moved either too quickly or too slowly. On the one hand, if the resistance is bypassed too quickly, the CEMF will not build up fast enough to limit current to a safe value. On the other hand, if we move the arm too slowly, a heavy starting current may flow long enough to damage the starter resistance. Consequently, the operator of a motor starting box must judge the movement of the wiper arm carefully.

The sound of a motor is usually the key to the proper operation of the starting box. As a motor starts and comes up to speed, it produces a low-pitched rumble which gradually increases to a high-pitched, whining noise. At full speed, the noise usually disappears because the pitch rises until it is above the range of the human ear. When a starting box is used, the noise will increase in pitch until the speed limit for a particular starting position is reached. Then, it becomes almost constant as long as the motor remains at that speed. Thus, the operator holds the arm at each position until the pitch of the motor noise is no longer increasing; then he moves the arm to the next position.

All resistance starters operate on the same basic principle. They may vary greatly in mechanical design and physical appearance, but electrically they are the same. They simply provide a method of inserting, and then gradually decreasing, a resistance in the armature circuit of the motor.

12.5   AC Motors—General

Due to the widespread distribution of ac power, most motors you are likely to encounter will be of the ac type.

Single-phase motors are generally used for applications requiring 1 hp or less. Many types of single-phase motors are made and with different characteristics. Some have high starting torque, some are made for intermittent operation only, and still others are made with one major thought in mind, economy.

Above 1 hp, *three-phase* ac motors are much used. The most common type of motor in this general class is the *induction* motor.

12.6   Three-Phase Induction Motors

Induction motors are so called because current in the rotor is induced from the stator. The rotor has neither slip rings nor a commutator so it has no direct electrical connection to the supply. The stator windings are connected to the supply line and act like the primary of a transformer. The rotor windings act like the secondary windings of a transformer. When current flows in the stator windings, a voltage is induced in the rotor windings causing current to flow there, too.

In a typical three-phase induction motor, the frame supports the stator core. The core is made of laminated steel to cut down on eddy current and hysteresis losses. The stator windings are embedded in slots in the core.

The rotor core is also made of laminated steel. Heavy copper bars are embedded in slots on the core and the bar ends are welded to copper shorting rings. Because of its appearance, a rotor of this type is usually called a *squirrel-cage* rotor. The bars and shorting rings form short-circuited windings. When a voltage is induced in these windings, a large current flows through them.

Operation

The three-phase voltage applied to the stator of an induction motor sets up a rotating magnetic field. To see how this rotating magnetic field is produced, let us look at the connections to the stator.

Fig. 12-8 shows a delta-connected, two-pole, three-phase stator connected to a three-phase ac supply. (Two-pole means two poles for each phase.) The windings on poles 1 and 4 are supplied by phase A, the windings on poles 3 and 6 are supplied by phase B, and the windings on poles 2 and 5 are supplied by phase C. They are wound so that, when one pole of the pair is north, the other pole of the pair is south.

Recall that in a three-phase voltage each phase is displaced 120 degrees from the other two phases. As these voltages rise and fall they cause currents to flow in the stator windings. To see why the magnetic field will rotate, let us examine the magnetic fields at different times in one cycle.

Figure 12-9(a) shows the current waveform I_A through windings 1 and 4 (phase A of Fig. 12-8) for one cycle, and the magnitude and polarity of the field this current sets up between poles 1 and 4 at different times in the cycle.

Figure 12-9(b) shows the same thing for current waveform I_B, which

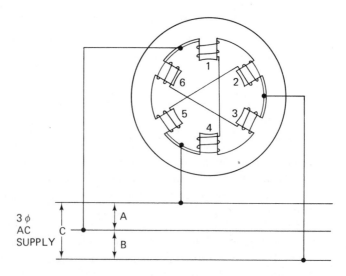

Figure 12-8   A delta-connected, two-pole, three-phase stator connected to a three-phase ac motor

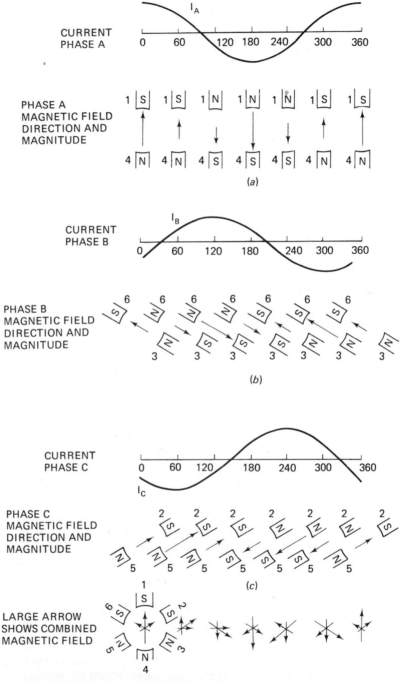

CURRENT
PHASE A

CURRENT
PHASE B

CURRENT
PHASE C

PHASE A
MAGNETIC FIELD
DIRECTION AND
MAGNITUDE

PHASE B
MAGNETIC FIELD
DIRECTION AND
MAGNITUDE

PHASE C
MAGNETIC FIELD
DIRECTION AND
MAGNITUDE

LARGE ARROW
SHOWS COMBINED
MAGNETIC FIELD

Figure 12-9   How a rotating magnetic field is produced by the stationary armature of a three-phase ac motor

lags waveform I_A by $120°$, and Fig. 12-9(c) shows the same thing for current I_C which lags current I_B by $120°$.

Figure 12-9(d) shows the combined field strength and direction for the three phases. The heavy black arrows represent the resultant magnetic field. As shown, the field rotates clockwise, and its strength is unchanged as it rotates.

Initially, current waveform I_A in Fig. 12-9(a) is maximum positive and the field between poles 1 and 4 is maximum, with 1 as the south pole and 4 as the north pole. At the same time, current I_B is a small negative value, setting up a small magentic field with 6 as the south pole and 3 as the north pole. Current I_C is also a small negative value, setting up a small magentic field with 2 as the south pole, and 5 as the north pole.

These three magnetic fields combine, as shown in Fig. 12-9(d), to form a single strong magnetic field with the south pole at 1 and the north pole at 4.

By examining the currents and fields through one complete cycle, you can see that the strength of the resultant magnetic field remains constant and that the field rotates one complete revolution for each cycle of the applied voltage.

Now let us see how this rotating magnetic field can be used to cause rotation of a squirrel-cage rotor which has many shorted conductors.

When the rotor is placed in the rotating field established by the stator, the lines of force cut the shorted conductors. This induces a voltage in the conductors and causes a large current to flow. The resulting magnetic field set up around the rotor is opposite to that surrounding the stator. Thus, the south pole is induced in the rotor opposite a north pole in the stator. As the stator field rotates, it attracts the rotor field, causing the rotor to turn in the same direction as the rotating stator field.

Because the rotor turns in the same direction as the rotating magnetic field of the stator, the direction of rotation can be changed by reversing the direction of field rotation which depends upon the phase relationships of the currents in the stator windings. Interchanging any *two* of the supply leads will change the phase relationship of two of the phases and cause the field to rotate in the opposite direction.

The rotor *does not* turn as fast as the rotating field of the stator. Let us see why. You know that the torque produced by a motor is proportional to the strength of the two fields that are interacting to cause rotation.

When the rotor is not turning, the stator field is cutting the rotor conductors at a maximum rate and, therefore, inducing the maximum voltage in them. When the rotor follows the rotating field and turns, less voltage is induced in the rotor conductors because the field cuts the conductors at a slower rate. Remember, the rotor and stator are turning the the same direction. If the rotor speed were to become equal to that of the stator field, no voltage would be induced in the rotor. With no voltage induced in the rotor, there would be no current flow, no field around the rotor, and no turning force, so the rotor would slow down. As soon as the rotor would slow down, of course, the lines of force of the stator field would once again cut the rotor conductors and create a new rotor field which would react with the stator field to produce rotation.

Synchronous speed

The speed at which the rotating stator field turns is called the *synchronous* speed of the motor.

With a 60 Hz supply voltage, the synchronous speed is 3600 rpm. The synchronous speed can be decreased by adding poles. Looking back at Fig. 12-8, suppose the six poles are moved around so that they occupy only half the circumference of the stator. One cycle of supply voltage will still cause the magnetic field to rotate past the six poles, but now this will be only a one-half turn. To get even torque we must, of course, fill in the other half of the stator with poles. This gives us two pairs of poles per phase, and the field will rotate at half speed. Using three paris of poles per phase will cut the speed to one-third.

The synchronous speed for any ac induction motor is expressed by the formula

$$N = \frac{f \times 120}{P} \qquad\qquad (12\text{-}2)$$

where N is the synchronous speed in rpm, f is the frequency of the applied voltage in Hertz, and P is the number of poles per phase. In an ac motor with four poles per phase and a 60 Hz supply, for example, the synchronous speed is

$$N = \frac{60 \times 120}{4} = 1800 \text{ rpm}$$

Slip and torque

As noted earlier, the rotor of an induction motor turns more slowly than the rotating field. The difference between the synchronous speed and the speed of the rotor is called slip. With more slip, more current flows in the rotor and the rotor develops more torque. Also, as the slip increases, the speed of the motor decreases.

Thus, motor speed is equal to synchronous speed minus slip. For example, a motor with a full-load speed of 1750 rpm and a synchronous speed of 1800 rpm has a slip of 50 rpm. Slip is expressed as a percentage of the synchronous speed. In this case,

$$\% \text{ slip} = \frac{1800 - 1750}{1800} \times 100$$

$$= 2.77\%$$

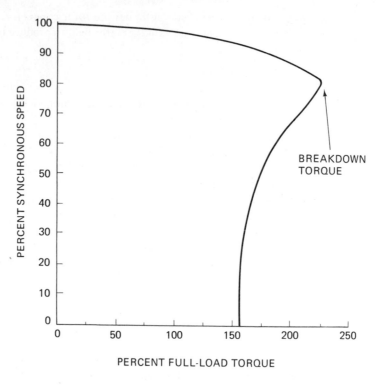

PERCENT SYNCHRONOUS SPEED

PERCENT FULL-LOAD TORQUE

Figure 12-10   Speed-torque curve for a three-phase squirrel-cage induction motor

Increasing the load on the motor will increase both slip and torque. Induction motors have very good torque at near synchronous speeds. Figure 12-10 shows the speed-torque curve for a typical three-phase induction motor. The percent of full-load torque is plotted against the percent of synchronous speed.

At no-load, the percent of full-load torque is nearly zero and the motor runs at nearly 100 percent of its synchronous speed. As the load is increased, the percent of full-load torque increases, and the speed falls off slightly. At full load (100 percent full-load torque) the slip is about 4 percent, so the motor runs at 96 percent of its synchronous speed. With the load increased so that the torque is 200 percent of full-load torque, the speed is still nearly 90 percent of the synchronous speed.

As the load is increased still further, the motor slows down and torque decreases. The point at which torque decreases as speed drops is called the *breakdown* torque. Stalling occurs when the speed drops to zero and torque is about 160 percent of the full-load torque. This is the starting torque of the rotor, also called *locked-rotor* torque.

It would appear that the motor should have its greatest torque when the rotor is not turning, because the motor is then drawing maximum current.

With the rotor stopped, however, there is a physical displacement of the rotor field in relation to the stator field. With a locked rotor, this displacement of the rotor field is enough so that part of the rotor field is developing a torque in the opposite direction. This subtracts from the total torque.

To see why this loss of torque takes place, we must examine what is called the *inductive reactance* of the rotor at different speeds. By definition, inductive reactance is the opposition of a coil to an alternating current. Mathematically,

$$X_L = 2\pi f L \qquad (12\text{-}3)$$

where X_L is the inductive reactance in ohms, π is a constant having the approximate value of 3.14, f is the frequency of the alternating current in Hz, and L is the inductance of the coil in henrys.

Now, recall that the field set up in the rotor is *in phase* with the rotor current. The rotor current *lags* the induced voltage by an amount dependent upon the inductive reactance of the rotor. As indicated by Equation 12-3, inductive reactance varies with frequency; the higher the frequency, the greater the reactance.

When the rotor is stopped, the voltage induced in the rotor is the same frequency as the line voltage. At the line frequency (60 Hz) the inductive reactance of the rotor is high enough to cause rotor current to lag the induced voltage by a large angle. The field set up around the rotor is in phase with this current. Thus, the rotor field is physically lagging the stator field far enough so that part of the rotor conductors are displaced into a field of the opposite polarity. These conductors produce a torque in the opposite direction and subtract from the total torque.

As the rotor speeds up, the frequency of the induced voltage, which is equal to the line frequency minus the rpm of the rotor, decreases.

If, for example, the synchronous speed of a motor with a 60 Hz supply is 3600 rpm (60 rps), and the slip is 4 percent, the rotor will be turning at 96 percent of the synchronous speed − 0.96 × 60 = 57.6 rps. The frequency of the induced voltage under these conditions is 60 − 57.6 = 2.4 Hz. The inductive reactance of the rotor winding at this frequency is negligible. The rotor field lags the stator field by a small angle and essentially no opposing torque is developed. The total torque of the motor is, therefore, increased.

12.7   *Two-Phase Induction Motors*

Although two-phase motors find very limited application, we will take a brief look at this type of induction motor for the sake of completeness.

The schematic diagram of a two-phase induction motor is shown in Fig. 12-11. The separate windings for each phase are displaced from one another by 90°. The rotor is the squirrel-cage type.

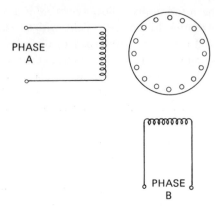

Figure 12-11   Schematic diagram of a two-phase motor

As shown in Fig. 12-11(b), poles 1 and 3 are energized by phase A, and poles 2 and 4 by phase B. The voltages for the two phases are 90° out of phase.

The magnetic fields formed by the supply voltage combine to form a single resultant field. Just as in the three-phase motor, the resultant magnetic field has a constant strength and rotates one complete revolution for each cycle of the supply voltage. Thus, the synchronous speed for a two-pole stator is 3600 rpm, as per Equation 12-2. With more poles per phase, the synchronous speed decreases.

The direction of rotation depends on the phase relation of the supply voltages. If phase A leads phase B by 90°, rotation is in a clockwise direction. If either phase A or phase B is shifted 180°, phase B will lead phase A by 90°, and rotation will be counter-clockwise.

The explanation on slip and torque given for the three-phase motor applies also to the two-phase motor.

12.8   *Other Polyphase Motors*

The two- and three-phase induction motors are the most common polyphase types. Other types are made, however, for special purposes. Some large three-phase motors have a wound rotor with slip rings and brushes connected to the external circuit. The wound rotor construction makes it possible to control the speed and torque of the motor by use of an external resistance.

Another special-purpose three-phase motor is the so called synchronous motor. In this motor, the rotor windings are supplied with dc. The dc produced field surrounds the rotor and follows the rotating magentic field around the stator. The rotor locks in step with the rotating field, and since there is no slip, the rotor turns at a constant synchronous speed. The rotors often have shorted squirrel-cage windings which allow the motor to operate as an induction motor until it comes up to speed.

12.9   *Single-Phase Motors*

By itself, a single-phase supply will not produce a rotating field in a motor. Instead, the flux field pulsates; that is, it builds up in one direction, falls to zero, and then builds up in the opposite direction. Under these conditions, the inertia of the rotor prevents rotation. If the rotor is mechanically turned until it is brought up to speed, however, it will then continue to rotate at near synchronous speed. The rotor will rotate in either direction, depending on how it is started.

To use single-phase ac for an induction motor, therefore, some means must be used to start the motor. This is accomplished by so called *phase splitting* which gives the *effect* of two phases and makes it possible to have a rotating field. Let us examine several methods of phase splitting.

Resistance-start motor

One means of splitting the phase of the supply voltage is to use a separate additional field winding during starting. Figure 12-12 shows a schematic diagram of the motor. There are two field windings and a squirrel-cage rotor. Notice that the main-field winding and the starting winding are spaced 90° apart like the windings in a two-phase motor.

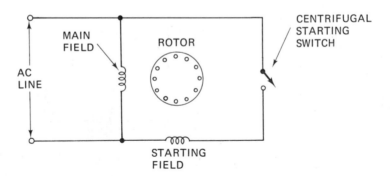

Figure 12-12   A split-phase motor

The main winding has very-low resistance and high inductance and, as a result, the current lags the line voltage by nearly 90°. The starting winding, on the other hand, has high resistance and low inductance so that its current is nearly in phase with the line voltage. This means there is a phase difference between the currents in the two windings. The difference is not a full 90° as in a two-phase motor, but the effect is the same, and a rotating field is generated.

As the motor builds up speed, a centrifugally operated switch opens the starting winding, and the motor continues to run as a single-phase induction motor.

Figure 12-13 shows one type of centrifugally operated switch. In Fig. 12-13(a), the motor is stopped and the switch is closed. The switch is held closed

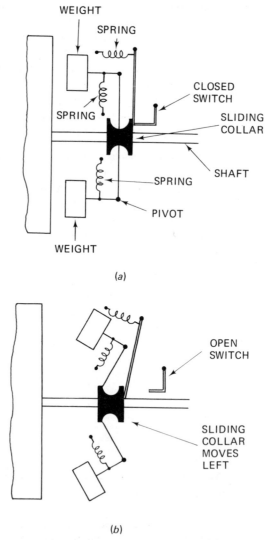

(a)

(b)

Figure 12-13   A centrifugal starting switch: (a) armature just starting to turn, and (b) armature rotating near full speed

by a pair of movable weights connected by pivoted levers to a movable collar on the shaft. As the shaft turns, centrifugal force causes the weights to move outward against spring tension. The pivoted levers push the collar to the left on the shaft and, as shown in Fig. 12-13(b), the switch opens and so disconnects the starting winding.

Figure 12-14 shows the effect on motor torque and speed when the switch opens. As the motor starts, the torque is about 140 percent of full-load torque. As the motor speed increases, the torque builds up. At the dashed horizontal line in Fig. 12-14 the switch opens and the motor operates as a single-phase induction motor. Notice the drop in torque.

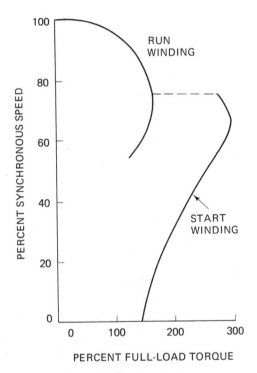

Figure 12-14   Speed-torque curve for a split-phase, resistance-start induction motor

The slip for a given load is greater than for polyphase motors. With no load, the speed approaches synchronous speed.

Capacitor-start motor

In the resistance-start motor, the phase difference between the currents in the two windings is not a full 90°. We can make the phase angle nearer 90° by using a capacitor to shift the phase in the starting winding.

The schematic diagram of a capacitor-start motor is shown in Fig. 12-15(a). This motor also has two windings spaced 90° apart and uses a squirrel-cage rotor. The centrifugal switch connects the starting winding to the line through a capacitor during starting. Current in the main winding lags the applied voltage. The inductive reactance of the starting winding is canceled by the opposite reactance of the capacitor to keep the current through the starting winding in phase with the applied voltage. Thus, the current in the main winding lags the current in the starting winding by 90° and the motor acts like a two-phase motor.

The speed-torque curve in Fig. 12-15(b) shows much higher starting

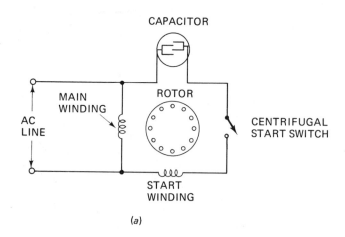

(a)

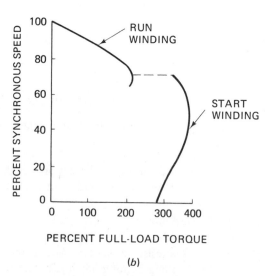

(b)

Figure 12-15   Capacitor-start motor: (a) schematic diagram and (b) speed-torque curve

torque than for the resistance-start motor. The dashed horizontal line shows that the torque changes when the starting winding is cut out. Now the motor comes up to speed as a single-phase induction motor. A large capacitor and a low resistance starting winding are used so that the starting winding draws heavy current to obtain a high starting torque. The motor would overheat if the large capacitor were left in the circuit at all times. If a smaller capacitor and a higher-resistance winding are used, however, the centrifugal switch can be omitted. The capacitor is then left in the circuit at all times and overheating does not occur. Under these conditions, the motor will not produce as much torque, but it will still be self starting.

The capacitor-start motor is probably the most frequently employed of the single-phase motors. They are made with many variations, but all employ the basic principle described here.

Split-phase, reversible motor

Like the capacitor-start motor, the split-phase, reversible motor uses a capacitor to shift the phase. Refer to the schematic diagram of Fig. 12-16. Two leads are brought out to a reversing switch. With the switch up, field A is the main winding, and field B is supplied through the capacitor. Then, the current in field A lags that in field B by a large angle. With the reversing switch moved down, field B becomes the main winding and field A is supplied through the capacitor. Then field B lags field A and the motor turns counterclockwise.

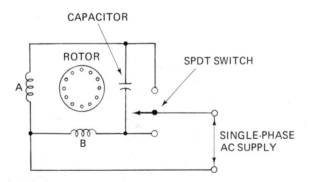

Figure 12-16   A split-phase reversible induction motor

Shaded-pole motor

Another type of small ac induction motor that is widely used is the so called shaded-pole motor. See the schematic diagram of Fig. 12-17(a). The motor has a squirrel-cage rotor and a single-field winding. The two-pole pieces are split, and a copper ring is fastened around one part of each pole piece. This part is called the shaded pole. The copper ring acts like the secondary winding of a transformer. As the current in the field winding builds up on one half-cycle, a flux builds up in the poles. This changing flux cuts the copper ring and induces a current in it. The current in the copper ring then sets up a flux which opposes the flux producing it. The main pole has a strong field, while that of the shaded pole is relatively weak.

When current in the field winding decreases, the main-pole field also decreases but current continues to flow in the copper ring. The shaded pole now has a strong field and the main-pole field is decreasing. When the main-pole field is zero, the shaded-pole field is maximum. Now the main-pole field reverses direction as current through the field winding reverses. The shaded-pole field changes later than the main-pole field.

In effect, the field in the shaded-pole section lags behind that of the main pole by nearly 90°. This gives the effect of a rotating field and produces rotation of the rotor. The rotor turns toward the shaded-pole section. In Fig. 12-17 the direction of rotation is clockwise.

As shown in the speed-torque curve of Fig. 12-17, the shaded-pole motor develops very low starting torque, has high losses, and cannot be reversed. Its top speed with no-load is less than any of the other single-phase motors. This indicates high losses. Low initial cost makes the shaded-pole motor popular, however, in applications where no-load starting is permissible. Fan motors, for example, is one very popular application.

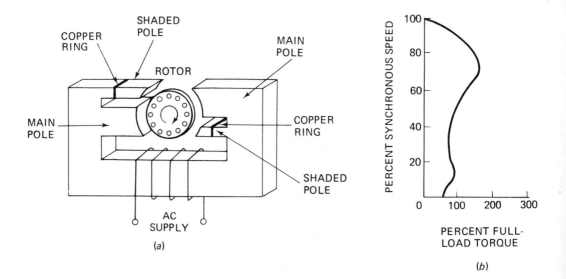

Figure 12-17   Shaded-pole motor: (a) schematic diagram and (b) speed-torque curve

12.10   *Single-Phase Motors With Brushes*

General

Several types of ac motors use commutators and brushes like those found on dc motors. These ac motors all have higher torque characteristics than the single-phase induction motors. They are made for many special purposes where small size and high starting torque are needed and the only power supply available is ac.

Universal series motor

This motor has a commutator and brushes and operates from either an ac or dc supply. A schematic diagram of the motor is shown in Fig. 12-18. Note that it is exactly like the series dc motor. Its principle of operation on ac is the same as on dc. Recall that reversing both the field and armature currents reverses the direction of rotation of a dc motor. With an ac supply, the current through both the field and armature reverses 120 times a second. While the torque drops to zero 120 times a second, the torque is always in the same direction.

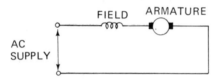

Figure 12-18   A universal series motor

The big advantage of the universal motor is *not* that it can be operated on either ac or dc. The universal series motor has about the highest starting torque for its size of any ac motor. Its speed-torque characteristics are similar to those of the series dc motor. This makes it extremely useful for applications in saws, portable drills, and similar devices.

With a large bit on a power drill offering a heavy load, for example, the motor will slow down and produce large torque. The same drill using a small bit with a light load will speed up. Thus, both drill bits are operated close to their best cutting speed.

Universal series motors are made in many specialized ratings, so a wide range of full-load operating speeds is possible.

Two-field reversible motor

A variation of the universal series motor is the two-field reversible motor shown in Fig. 12-19. Two separate field windings are wound on the stator in opposite directions. When one winding is energized, the motor turns clockwise and, when the other winding is energized, the motor turns in a counterclockwise direction. A lead from each field is brought out to a SPDT switch. One field winding is energized at a time to give the desired direction of rotation. In all other ways, this motor is like the universal series motor.

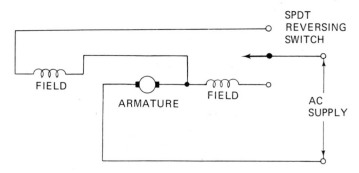

Figure 12-19   A two-field reversible motor

Repulsion motor

This is another small ac motor having high starting torque. Like the universal series motor, it has a commutator and brushes. Its principle of operation, however, is somewhat different. Refer to the schematic diagram in Fig. 12-20. The stator winding is connected directly across the ac line. There is no electrical connection to the rotor or brushes and the brushes are shorted together. The magnetic field of the stator induces a voltage in the rotor and the shorted brushes allow a heavy current flow in the rotor windings. This current sets up a field in the rotor which reacts with the stator field to produce rotation.

The position of the brushes is important in determining the direction of rotation and the speed and torque of the motor. When the brushes are in line with the flux of the field, they are said to be in their *hard neutral* position. A high current will then flow in the rotor windings but the motor will not turn because the fields are in opposite directions.

If the brushes are shifted 90° so that they are at right angles to the field flux, there will be no current flow in the rotor windings. This position is called *soft neutral* and, again, the rotor will not turn. To develop torque in the repulsion motor, the brushes must be set between hard neutral and soft neutral. Then there is current flow in the rotor windings and the rotor field is in phase with the stator field. Commercial motors of this type usually operate with the brushes set 15° to 20° from hard neutral. Moving the brushes to the opposite side of hard neutral will reverse the direction of motor rotation.

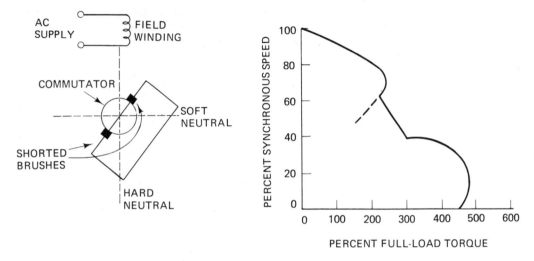

Figure 12-20   Repulsion motor: schematic diagram (left) and speed-torque curve (right)

Some motors of this type start as repulsion motors and, after they are up to speed, operate as induction motors. A centrifugal governor operates a switch to short the commutator segments when the motor is up to speed. The rotor then acts like a squirrel-cage rotor.

The speed-torque curves for a typical repulsion-start motor are shown in Fig. 12-20. Note the extremely high starting torque for this motor—over 400 percent of full-load torque. In the figure, note also at the dashed line on the curve, that the commutator segments are shorted and the motor comes up to speed as an induction motor. These motors are used where very high starting torque is needed.

Single-phase synchronous motors

Some synchronous motors operate from a single-phase supply. Like the three-phase synchronous motor, their speed is constant. These motors are widely used in electric clocks and other types of timing devices.

Many different kinds of small synchronous motors are designed to operate on single-phase ac. The exact mechanical arrangement varies widely, but they all depend on the same basic principle for operation at synchronous speeds.

The shaded-pole motor of Fig. 12-17(a) will operate at synchronous speed if its squirrel-cage induction rotor is replaced by a rotor of hardened steel which tends to retain the magnetism induced in it. The rotating field induces a field in the hardened-steel rotor. The reaction between the rotating field and the field induced in the rotor produces torque that causes the rotor to turn. The motor starts in the same way as an induction motor. As the motor approaches synchronous speed, the magnetic properties of the core tend to produce permanently magnetized poles along some fixed diameter of the rotor. When this happens, the rotor locks in step with the rotating field and rotates at synchronous speed. This motor is self-starting and is synchronous over a limited torque range. Its speed is 3600 rpm when opeated on 60 Hz current.

Figure 12-21 shows another type of small single-phase, synchronous motor. The speed of this motor is locked to the supply frequency, but as it runs at a sub-multiple of the synchronous speed, it is called a sub-synchronous motor. This motor is also self-starting and has the advantage of running at slow speed.

The speed of the motor depends on the number of teeth on the rotor. The rotor moves a pair of teeth past the poles for each cycle of ac. A motor with 16 teeth will, therefore, run at 450 rpm with a 60 Hz supply voltage.

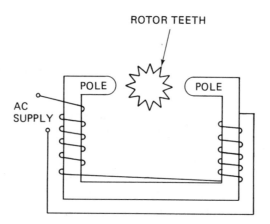

Figure 12-21   A sixteen-pole, sub-synchronous motor

12.11   *General Care and Maintenance*

Preventive maintenance

Preventive maintenance includes such jobs as testing, cleaning, drying, oiling, and adjusting motors. For best results, each motor should have its own history card which may either be attached to the machine or kept in a separate file. The card lists such items as the name of the manufacturer, model number, serial number, dates of purchase and installation, and the dates of inspection, brush changes, or overhaul.

In doing any work on motors, refer to the manufacturer's literature for operating limits, types of lubricants, replacement parts, and so forth.

Service factors

Formerly, motors were rated on the basis of temperature rise *above* a specified ambient temperature. The amount of temperature rise was generally limited to 40°C (72°F).

Today, motors have a so called *service factor* stamped on their name plates. This factor ranges from 1.00 to 1.35. A rating of 1.00 means that, when the motor is installed where the ambient temperature is not over 40°C (104°F), it can safely deliver its rated horsepower continuously. When the service factor is greater than 1.00, multiply the rated horsepower by the service factor to determine the horsepower the motor can safely deliver on a continuous basis with the same temperature limitation. If, for example, a 2 hp motor has a service factor of 1.15 (most common for motors over 1 hp), it can safely and continuously deliver 1.15 × 2 = 2.3 hp, provided the ambient temperature does not exceed 40°C (104°F). At higher temperatures, continuous operation at the full-rated horsepower will seriously limit the life of the motor.

The service-factor rating of a motor should not be confused with the so-called "overload capacity" of almost any motor. This simply means that the motor can deliver, for relatively short periods of time, 1.5 to 2.0 times its normal horsepower without harm. Consistent overload operation is a very poor practice, but when the overload does not occur frequently, the additional expense of securing a motor with a larger horsepower rating may not be justified.

Resistance readings

The resistance between the field or armature winding and the frame should be a very high value, usually measured with an instrument called a megohmmeter. This measurement should be taken with the power line completely disconnected and should be recorded on the history card. The resistance will vary from time to time due to variations in temperature, the amount of moisture between the windings, and foreign materials, such as dust, on the windings. The best time to take resistance readings is immediately after the machine has been shut down so that the heat has dried out the moisture. Moisture causes the resistance to read lower than normal.

A large change in a motor's resistance reading may occur over a period of months. This is an indication of trouble which may be caused by the accumulation of dirt, grease, or moisture in the insulation of the windings. If the windings are not cleaned and dried out, the lowered resistance will cause leakage currents, which will eventually destroy the insulation and cause short circuits.

Cleaning methods

An air hose, a vacuum cleaner, or a clean unused paint brush will usually remove surface dust or dirt. Where dirt is encrusted on the windings, you will have to scrape it off with a wooden or plastic paddle. *Never* use metal or anything with a sharp edge, since either might damage the insulation.

Oil and grease can be removed by wiping the insulation with a piece of cheesecloth dampened with an approved petroleum solvent such as Stoddard solvent. Carbon tetrachloride should *never* be used because it will cause deterioration of the rubber insulation and commutator film and, furthermore, its fumes are poisonous.

A safe method of cleaning electrical machinery is to use fresh water and household detergent. The winding should *not* be soaked in the water. Apply the water with a scrub brush. This should be done as quickly as possible and the windings should then be wiped dry. Before reassembling the machine, it should be baked so that all moisture is removed. If the machine is small, it can be baked by placing it on top of an oven, steam pipes, or a radiator. Larger units should be placed in an oven or a heat chamber.

Commutators

Commutators acquire a film after a week or so of operation. This film is made up of particles from the commutator, carbon particles from the brushes, and moisture from the air. The film may range in color from very light tan to almost black. The best color for good contact and minimum wear is about the color of milk chocolate.

The film, which is the result of friction between copper and carbon, has a definite advantage. It reduces wear, acts as a lubricant, and tends to reduce arcing at the brushes.

To keep the proper composition of this film, all replacement brushes must be the exact type recommended by the manufacturer.

Commutators should be cleaned often enough to prevent the surface from becoming oily or greasy. Hand-woven duck canvas is a suitable cleaning material. Press the canvas against the rotating commutator to wipe it efficiently. *Never* use solvents or any other item except the canvas to clean the commutator.

The color and uniformity of the film on the commutator indicates its condition. Look also to see if the mica between commutator segments is properly undercut. If the mica is high, it will cause the brushes to ride up and arcing will take place between the brushes and the commutator segments. Arcing will soon pit the copper and cause excessive wear. In extreme cases, the commutator will have to be resurfaced by a shop specializing in motor overhaul.

Brushes

The brushes are critical parts of any motor. They are fitted into a brush holder and rest against the commutator. The pressure of the brush is governed by a spring which presses against the outside end of the brush. Brushes will give very little trouble when they are the correct type and the correct pressure is applied.

When a new brush is installed, it should be shaped to the commutator. One way to do this is to put a piece of sandpaper between the commutator and the brush, with the sanded surface pressing against the brush. Rotate the commutator back and forth by hand until the face of the brush assumes the contour of the commutator.

Bearings and lubrication

In small motors the bearing may be a bronze sleeve with grooves to allow oil to reach the shaft. Larger motors have ball bearings. Some motors also have an additional bearing, called a thrust bearing. In may be a single steel ball between the end of the shaft and the housing.

Never lubricate a motor without first referring to the manufacturer's literature. *Much damage can be done by over-lubrication.* Moreover, an excess of oil and grease decreases the resistance of the insulating material, collects dirt, and causes fire hazards.

Indications of faulty bearings are noisy operation, overheating of bearings, or freezing of the shaft. Clearance between rotors and stators is very small. A bad bearing can cause a rotor to rub against the field poles.

Unless you have specialized training in motor overhaul, confine your work on motors to their installation and routine maintenance as described here.

12.12   *Motor Wiring*

It is unlikely that, until you have gained considerable experience in wiring, you will be called upon to design a motor-circuit installation. Instead, it is much more likely that you will assist in the installation of a motor circuit designed by an experienced electrical contractor who is familiar with the various ordinances governing such work. It is improtant, however, that you gain from this text some understanding of the factors involved in such jobs.

Because we are not concerned with industrial applications in this text, our discussion of motors in this section is concerned primarily with overcurrent protection, how a motor is totally disconnected from a circuit, and the selection of wire sizes. Moreover, we will concern ourselves solely with single-phase ac motors since this is the type you will most often encounter. For more detailed information on specific motor installations, refer to Code Article 430—Motors, Motor Circuits, and Controllers.

Motor circuits—general

When a single motor is connected to a branch circuit, four components are required: **(a)** an overcurrent protection device, **(b)** a controller to start and stop the motor, (3) running-overcurrent protection, and (4) a disconnecting means to totally disconnect the motor and controller from the circuit. Quite often, more than one of these components are combined in a single device.

Overcurrent protection

Code Ref. 430-42(d) reads as follows: "The overcurrent device protecting a branch circuit to which a motor or motor-operated appliance is connected shall have sufficient time delay to permit the motor to start and accelerate its load." Since the starting current of a motor is much heavier than its running current, this reference indicates that either a thermal-type circuit breaker or a time-delay fuse should be used for overcurrent protection. The amperage of the overcurrent device is, of course, determined by the size of the wire used in the branch circuit.

Controllers

On equipment that starts and stops automatically, such as a refrigerator, for example, the manufacturer provides the proper controller for the equipment. On manually-operated controllers, however, the selection of the controller is left to the electrician.

For portable motors rated at one-third horsepower or less, the cord and the plug are the controller.

For motors rated at 2 hp or less, a general use ac toggle switch having a rating of not less than 125 percent of the ampere rating of the motor is satisfactory. Alternatively, an enclosed knife switch may be used. If the switch is rated in amperes, the rating must be no less than two times the ampere rating of the motor. If the switch is rated in horsepower, the rating must be no less than that of the motor.

For motors larger than 2 hp, the switch must be rated in horsepower and the rating must be at least equal to that of the motor.

Special motor starters, which combine start-stop control and overload protection are also available. They are rated in horsepower and are operated either by a toggle switch or a pushbutton. On pushbutton types, the button may be mounted directly on the device enclosure or at some remote location.

A circuit breaker, rated in amperes only, may be used as a controller provided it meets the requirements for motor running-overcurrent protection. They are given below.

Running-overcurrent protection

The term "running overcurrent," as used here, is synonymous with the term "overload." Notice also that overload is an operating overcurrent which,

---

*   Reprinted with permission from the 1975 National Electrical Code, Copyright 1975, National Fire Protection Association, Boston, Ma.

when it persists for a sufficient length of time, causes damage or dangerous overheating of the apparatus. It does not include short-circuits or ground faults. Because the starting current of a motor is generally greater than a running overcurrent, the fuse or circuit breaker used for overcurrent protection on a branch circuit will not protect a motor from overload. Thus, a separate device is needed to provide that kind of protection.

Manually controlled, 115 volt portable motors of 1 hp or less may be plugged into ordinary 15 or 20 A branch circuits. Running overcurrent protection is usually—but not always—provided by a so called "thermal protector" which is integral with the motor circuit. The notation "Thermally Protected" will be found on the name plate of such motors.

As the name implies, a thermal protector is simply a heat-sensitive device which acts to open the circuit when the motor temperature exceeds a safe value. Once tripped, the thermal protector usually requires manual resetting.

Permanently installed motors not equipped with an integral thermal protector require separate running overcurrent protection. Most often, this device is a *motor starter* containing "heater coils" which match the amperage of the motor. Such starters carry the normal running current indefinitely and a small overload for some time, but trip on heavy overload or when the motor fails to start. The starter then acts as a combined controller and running-overcurrent protector.

If a separate circuit breaker is used for overload protection, it must be rated at not more than 125 percent of the motor amperage. The circuit breaker then acts as a combined controller, overload protector, and disconnecting means.

When a knife switch combined with a fuse is used for overload protection it must be rated at not more than 125 percent of the motor amperage. The fuse or fuses must, of course, be of the time-delay variety to prevent *blowing* when the motor starts. The switch-fuse combination also combines the functions of controller, overload protection, and disconnecting means.

Disconnecting means

Regardless of size, the plug and receptacle serve as the disconnecting means for *portable* motors. So the explanations which follow apply only to permanently installed motors.

No separate disconnecting means is required for motors rated at one-eighth horsepower or less. The branch circuit overcurrent device then serves as the disconnecting means.

A knife switch-fuse combination may serve as the disconnecting means on larger motors.

Motors having an integral thermal protector, or those employing a combined controller-overload protector, require a separate disconnecting means. An enclosed knife switch-fuse combination, with a horsepower rating not less than that of the motor, is often used for this purpose, even though the Code does not require the fuse inasmuch as overload protection is already provided.

Motors rated at two horsepower or less may also use a knife switch-fuse combination rated in amperes, *provided* the rating is at least two times the motor amperage. Alternatively, you may use a general purpose ac toggle switch rated at 125 percent or greater of the motor amperage. For either arrangement, an SPST switch is used for 115 volts and a DPST switch for 230 volts.

Wire sizes

The wires used with any motor must have an ampacity rating of no less than 125 percent of the motor nameplate current rating.

EXERCISES

12-1/   Explain what happens to a current-carrying conductor in a magnetic field.

12-2/   How is rotation of a motor shaft produced?

12-3/   What is meant by the expression *torque*?

12-4/   What is CEMF and how is it produced?

12-5/   What regulates the amount of CEMF produced by a motor?

12-6/   What is the *neutral plane* of a motor?

12-7/   What is meant by *armature reaction* and how does it affect a motor?

12-8/   What measures are taken in practical motors to counteract the effects of armature reaction?

12-9/   When does a motor produce its maximum torque?

12-10/   Define horsepower.

12-11/   What is the electrical equivalent of mechanical horsepower?

12-12/   Does a motor with 746 watts electrical-power input produce 1 hp at the output shaft? Explain your answer.

12-13/   If a motor runs at 1500 rpm at no-load, and at 1460 rpm with its rated load, what is its speed regulation?

12-14/   Describe a series motor and draw its schematic diagram.

12-15/   What are the speed-torque characteristics of a series motor?

12-16/   Can a series motor be run without a load? Why?

12-17/   Describe a shunt motor and draw its schematic diagram.

12-18/   What are the speed-torque characteristics of a shunt motor?

12-19/   How may the speed of a shunt motor be controlled?

12-20/   Describe a compound motor and draw its schematic diagram.

12-21/   What are the speed-torque characteristics of a compound motor?

12-22/   What is the difference between a *cumulative-compound* and a *differential-compound* motor?

12-23/   Draw the schematic diagram of a starting box intended for manual operation of a shunt-type dc motor and explain its operation.

12-24/   Explain the operation of a three-phase induction motor.

12-25/   In a three-phase induction motor, the rotor does not turn as fast as the rotating field of the stator. Why not?

12-26/   What is meant by the *synchronous* speed of a motor?

12-27/   What formula is used to determine the synchronous speed of any ac induction motor?

12-28/   What is meant by the *slip* of an induction motor and what formula may be used to determine the slip of a given motor?

12-29/   What are the speed-torque characteristics of a typical three-phase induction motor?

12-30/   How does inductive reactance affect the torque of an induction motor?

12-31/   Explain the operation of a two-phase induction motor.

12-32/   What is a synchronous three-phase motor?

12-33/   By itself, a single-phase supply will not produce a rotating field in a motor. What general technique is used, therefore, to start a motor?

12-34/   Describe the operation of a resistance-start, single-phase motor.

12-35/   Explain the operation of a centrifugal-start switch.

12-36/   How does the slip of a single-phase, resistance-start motor compare with that of a polyphase motor?

12-37/   Describe the operation of a capacitor-start motor.

12-38/   Describe the operation of a split-phase reversible motor.

12-39/   Describe the operation of a shaded-pole motor.

12-40/   Describe the operation of a universal series motor.

12-41/   Describe the operation of a repulsion-type motor.

12-42/   In a repulsion-type motor what is meant by the expressions *hard-neutral* and *soft-neutral*?

12-43/   Describe how a shaded-pole motor can be made to operate at synchronous speed.

12-44/   What is meant by the *service factor* of a motor?

12-45/   What conditions may cause a large change in the resistance readings of motor windings?

12-46/   Explain the methods for cleaning electrical machinery.

12-47/   Describe the general appearance and maintenance of commutators.

12-48/   What precautions should be taken when new brushes are installed?

12-49/   Explain the general provisions for overcurrent protection in motor circuits.

12-50/   Explain the general provisions for controllers, running overcurrent protection, and disconnecting means in motor circuits.

13
Farm Installations

13.1 *General*

In addition to the electrical devices and appliances found in city residences, farms require such things as milking machines, hay dryers, silo fillers, milk coolers, and special hot-water heaters for the dairy and chicken house. Consequently, the electrical power consumption on a farm is usually much greater than for a city residence. Moreover, the wiring system is spread out over much greater distances and may be a combination of overhead and underground.

Since the need for additional appliances and machines changes as a farm grows, the initial planning should provide for the liberal addition of circuits. The cost of installing additional capacity that is not used immediately is generally less than the cost of any major rewiring job, particularly in these days of sharply rising prices.

To avoid the problems associated with icing, lightning, and the movement of tall machinery commonly used on farms, an underground system should prove more satisfactory than an overhead system. The specifications in Sec. 9.9 of this text for residential underground services apply equally well to a farm.

On most farms, the service drop terminates on a meter pole in the farmyard. Separate feeders then run to the residence, barn, chicken house, repair shop, storage areas, and so on.

In each building there is a service entrance, but no meter. A ground is required for any building that contains more than one circuit or that houses livestock.

13.2   Grounds—General

Inasmuch as municipal water systems are not generally available to farms, "made-electrodes" must be used for grounding.

In all such cases the grounding electrode shall consist of a driven pipe, driven rod, buried plate or other device approved for the purpose and conforming to the following requirements: (NEC Sec. 250-83).

- **Concrete-encased electrodes.** Not less than (1) 20 feet of bare copper conductor not smaller than No. 4, or (2) steel reinforcing bar or rod encased by at least 2 inches of concrete and located within and near the bottom of a concrete foundation footing that is in direct contact with the earth.

- **Plate electrodes.** Each plate electrode shall present not less than 2 square feet of surface to exterior soil. Electrodes of iron, or steel plates shall be at least 1/4 inch thick. Electrodes of nonferrous metal shall be at least 0.06-in. thick.

- **Pipe electrodes.** Electrodes of pipe or conduit shall be not smaller than 3/4 inch trade size and, where of iron or steel, shall have the outer surface galvanized or otherwise metal-coated for corrosion protection.

- **Rod electrodes.** Electrodes of rods of steel or iron shall be at least 5/8 inch in diameter. Nonferrous rods or their equivalent shall be listed and shall be no less than 1/2 inch in diameter.

- **Installation.** Where practicable, made electrodes should be imbedded below permanent moisture level. Where rock bottom is not encountered, the electrode shall be driven to a depth of 8 feet. Where rock bottom is encountered at a depth of less than 4 feet, electrodes not less than 8 feet long shall be buried in a trench. All electrodes shall be free from nonconductive coatings, such as paint or enamel.

13.3   Ground Resistance

NEC Section 250-84 specifies that the resistance to ground of a made-electrode shall not exceed 25 ohms. Where the resistance to ground of a single made-electrode is greater than 25 ohms, two or more made-electrodes are connected in parallel.

The easiest method to determine the resistance to ground of a

---

made-electrode is by means of a ground-resistance meter. Because these instruments are quite expensive, however, and not always available, let us discuss an alternate method which provides a reasonable approximation.

Take an ac ammeter with a 0-30 A scale and mount it in a suitable box. Connect one meter terminal to a 30 A fuse mounted in the same box. From the opposite end of the fuse and from the remaining ammeter terminal bring out test leads with insulated clips at their ends. Use insulated AWG #12 wire for the test leads. Study this arrangement which is shown schematically in Fig. 13-1.

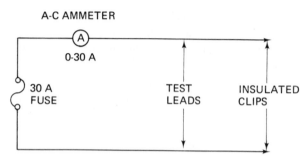

Figure 13-1   Arrangement of ammeter, fuse, and test leads to form a resistance-to-ground tester

Temporarily disconnect the grounding electrode conductor from the made-electrode. Now connect the ammeter assembly between the made-electrode and the "hot" conductor of the wiring system. This grounds the "hot" wire and the resulting fault current passes through the series-connected ammeter and fuse. Neglecting the resistance of the meter, the resistance of the ground can be determined by Ohm's law. If, for example, the line voltage is 120 V and the ammeter reads 6 A, the resistance to ground is

$$R = \frac{E}{I} = \frac{120}{6} = 20 \text{ ohms}$$

This is an acceptable ground, but you should try to improve it. Of course, the "hot" wire should be grounded no longer than necessary to obtain the required meter reading. Also, once the meter reading is obtained, make sure that you reconnect the grounding-electrode conductor between the system neutral wire and the made-electrode.

Generally, the hot wire will be protected by a 15 A fuse or circuit breaker at the service. In the example given, the fuse will not open, nor will the circuit breaker trip under a brief surge current of 20 A. If the service fuse blows or the breaker trips, immediately, you have an excellent low-resistance ground.

Is there any way to improve, that is reduce, the resistance to ground? Fortunately, there are several methods.

1.   Install the made-electrode where rain falling off the roof will tend to keep the ground around the electrode wet. The reason: the drier the ground, the higher its resistance.

2.   Drive a second rod 12 to 16 feet into the earth, at a distance of 10 feet or more from the first rod, and connect the two with ground clamps and wire of the same size of the ground wire. In extreme cases of too-high resistance, a third rod, spaced 10 feet from, and connected to, the other two may be required.

3.   Soak the ground around the made-electrode with a heavy concentration of salt water. Repeat this treatment at least once each year.

13.4   Meter Poles

As noted earlier, for most farms the service drop ends at a meter pole. To minimize expense, the meter pole should be located near the building with the greatest load. Smaller sized conductors are required for buildings with smaller loads.

A typical meter-pole installation is shown in Fig. 13-2. The service drop terminates at the top of the pole. Usually the neutral wire is the topmost wire, but this should be verified with the electric supplier. Also notice that the neutral is spliced to the neutral lines running to all buildings. In effect then, the neutral is continuous from the power line to each building. The neutral is also connected to the meter and is grounded at the pole. More is said about grounding later. Right now, let us see what happens to the "hot" wires.

The ungrounded conductors are run from the top of the meter pole to the meter socket, and from there, back to the top of the pole. Including the neutral wire, which grounds the meter socket for safety purposes, three conductors run down the pole and two conductors return to the top of the pole. Depending on local codes, all five conductors may be encased in a single conduit or, alternatively, separate conduits may be required for the "down" and "up" wires. When all wires are contained in one conduit, the assembly is called a "single stack." When two conduits are used, it is a "double stack." With either arrangement, leave at least one-third of the pole circumference clear to permit easy climbing.

When a main switch or circuit breaker is used, follow the wiring plan shown in Fig. 13-2. The switch or circuit breaker is weatherproof, of course, and is mounted about 5 feet above ground with the meter just above it. The two are joined by a nipple.

Insulator racks are installed near the top of the pole. Use one rack for each set of three wires and secure the racks to the pole by heavy through bolts which pass completely through the pole.

To solve the water problem, bring the top of the conduit to a point above the topmost insulator and place a service head at its top.

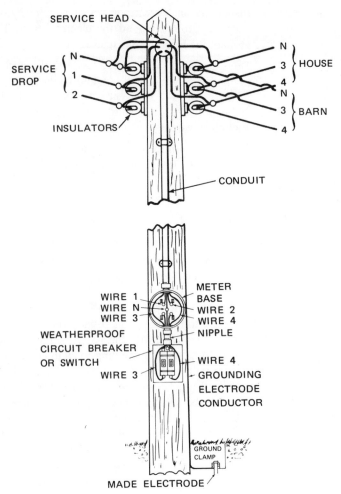

Figure 13-2   A typical meter pole

The switch, meter socket, wired conduit, and insulators are pre-assembled on the pole before it is placed in an upright position.

13.5   *Determining Feeder Loads*

The feeders that supply farm buildings (excluding family dwellings) or loads that consist of two or more branch circuits shall have minimum capacity computed in accordance with NEC Table 220-40, identified below as Table 13-1.

Load in amperes at 230 volts	Percent of connected load
Loads expected to operate without diversity, but not less than 125 percent full-load current of the largest motor and not less than first 60 amperes of load	100
Next 60 amperes of all other loads	50
Remainder of other load	25

Table 13-1   Method for computing farm loads for other than dwellings

The total load of the farm for service-entrance conductors and service equipment shall be computed in accordance with the farm dwelling load and demand factors specified in Table 13-1. Where there is equipment in two or more farm equipment buildings or for loads having the same function, such loads shall be computed in accordance with Table 13-1 and may be combined as a single load in Table 13-2 for computing the total load. (See NEC Sec. 230-21 for overhead conductors from a pole to a building or other structure.)

Individual loads computed in accordance with Table 13-1	Demand factor percent
Largest load	100
Second largest load	75
Third largest load	65
Remaining loads	50

Table 13-2   Method for computing total farm load

To illustrate the use of the above information, let us compute the load for a farm. Initially, we will exclude the residence.

Building 1: - Dairy Barn	
Lighting	2500 W
Water heater	3000 W
Total	5500 W
5500 W/230 V =	24 A
Milker	14 A
Cooler	8 A
Air conditioner	15 A
Total	61 A

Building 2: - Chicken House

Brooder	4000 W
Lighting	600 W
Total	4600 W
4600 W/230 V =	20 A

Building 3: - Storage and Repair

5 hp (230 V, single phase) motor	26 A
2 hp (230 V, single phase) motor	10 A
Lighting	8 A
Total	44 A

The largest motor employed is 5 hp. From Table 13-1, 125 percent of 26 A is 32.5 A. This increases the total load for building 3 from 44 to 50.5 A.

The total load for the three buildings is:

Building 1	61 A
Building 2	20 A
Building 3	51 A
Total	132 A

From Table 13-1:

First 60 A at 100%	60 A
Next 60 A at 50%	30 A
Remainder at 25%	3 A
Total	93 A

Now let us consider the same farm but include the residence. Assume the house has a floor area of 2000 square feet, excluding an unfinished attic. The range is rated at 11 kW.

Lighting	6000 W
Small appliances	3000 W
Laundry	1500 W
Total (excluding range)	10,500 W
3000 W at 100%	3000 W
7500 W at 35%	2625 W
Net load (excluding range)	5625 W
Range (NEC Table 220-5)	8000 W
Net load (including range)	13,625 W
13,625 W/230 V =	59 A

Largest demand, Bldg. 1 at 100%	61 A
Second largest demand - Residence at 75%	44 A
Third largest demand - Bldg. 3 at 65%	29 A
Balance of demand - Bldg. 2 at 50%	10 A
Total demand	144 A

13.6   *More About the Meter Pole*

When service amperage runs 200 A or higher, a current transformer is used to reduce the size of conductors used between the top of the meter pole and the meter proper.

Some current transformers have a single-turn primary in the form of a solid bar. This primary is connected in series with the load to be measured. In farm installations, however, a "doughnut" type current transformer is more common. The name derives from its resemblance to that familiar pastry.

The doughnut transformer has a single winding—the secondary. The "hot" conductors whose current is to be measured pass through the center of the doughnut and act as the primary. The secondary is magnetized by the magnetic lines surrounding the primary. The secondary has many turns of small wire and is usually connected to a 5 A ammeter. The ratio of turns between the two circuits depends on the anticipated magnitude of the current to be measured. If, for example, the anticipated current is 200 A, the actual turns ratio would be 40:1. It is customary, however, to refer the ratio to the full-scale reading of the ammeter. Thus the ratio is designated 200:5.

Because it consumes very little power, a current transformer is a fairly small device, typically measuring about 5 inches in diameter and just slightly more in length. A mounting bracket is attached to the bottom of the transformer case and the secondary terminals extend through the top of the case.

The ammeter connected across the secondary effectively forms a short circuit. When the ammeter is not connected, a high voltage is induced in the secondary because the primary load current is not dependent on the secondary load current. The primary then acts as an inductance and this creates a voltage drop in the primary. Due to the high secondary-to-primary turns ratio, the resulting open-circuit voltage of the secondary may reach thousands of volts. To remove this safety hazard, a switch is connected in parallel with the secondary and with the ammeter and is closed when the ammeter is not in use. Appropriately, this switch is called a "shorting" switch.

The current transformer and shorting switch are usually installed in the same cabinet which is located near the top of the meter pole, just under the service drop insulator bracket. The switch is operated from the ground by means of an extension handle secured to the pole and terminating about 5 feet above ground.

Two AWG #14 wires run from the transformer secondary to the current winding of the watthour meter. Two additional AWG #14 conductors run from "hot" wire terminals inside the switch cabinet to the voltage winding of the meter.

For the best protection against lightning, run the neutral wire directly from the top of the pole (near the service drop termination of the neutral) to the ground rod at the bottom of the pole. Depending on local preference, the neutral lead may be stapled to the pole on the side opposite from the conduit or it may be tucked in alongside the conduit as far as it runs, then run to ground.

The ground rod is typically located about 2 feet from the bottom of the pole with a 1 foot deep trench between the rod and pole. The neutral runs from the base of the pole, through the trench, and is attached to the rod by means of a ground clamp. The entire assembly is then covered.

Avoid driving ground rods in locations saturated by animal excreta since the chemical breakdown of such degradable matter may destroy all or portions of the ground assembly.

13.7   *Building Entrances*

A building wired with nonmetallic-sheathed cable and having only one circuit need not be grounded. But grounding is mandatory wherever any metal equipment (switch boxes, conduit, etc.) is used.

At its point of entry to the building run the grounding electrode-conductor between the neutral and the grounding electrode.

The residence will contain either circuit breakers or a fused-service switch in conjunction with branch-circuit fuses. For convenience, circuit breakers are recommended.

In all buildings excluding the residence, one or more switches must be provided to disconnect all wiring. When buildings are fed current from another building, the switch or circuit breaker which controls the wiring may be located either at the building of origination or at the building of termination. When the fuse at the starting point protects the smallest wire in the entire circuit, the switches in other buildings need not be fused. Conversely, if the fuse at the starting point does not protect the smallest wiring in the circuit, an appropriate fuse is required at the point where the smaller wire is tapped off the larger.

Because good grounds are often difficult to locate on farms, conduit and armored cable are rare; therefore, the NEC recommends nonmetallic sheathed cable. To make the entire system nonmetallic, and thereby improve safety, use nonmetallic outlet and switch boxes made of various plastic materials. These boxes are installed the same as the metal variety, but the nonmetallic sheathed cable must be supported within 8 inches of a box. Since the nonmetallic boxes contain no cable clamp, one additional conductor may be installed in the box.

13.8   *Miscellany*

In buildings that house livestock, you must use types NMC or UF cable. Of course, either type is suitable for the entire installation, but there may be occasions when special circumstances dictate some other choice. Always check on the prevailing code.

Locate switches so that they can be operated by a shoulder or elbow. A farmer's hands are often full when he enters a building.

For the same reason that a nonmetallic system is recommended, use only procelain or bakelite sockets.

For physical protection, locate lamp outlets between joists so that the lamps do not project too far into closely confined spaces, such as aisles.

For improved lighting, use a reflector with every socket, and clean the reflectors at frequent intervals.

Local codes may make vaporproof receptacles mandatory in the vicinity of haymows.

When cable is located where the possibility of damage exists, make sure that it is provided with some form of protective covering. When the cable passes through a floor, it must be protected by at least a 6″ conduit or pipe.

Keep exposed cables away from the centers of aisles. Even though more cable may be required, do not run it along the bottoms of joists; instead, run it along the side of a joist.

To ensure operation of a water pump, particularly during periods of emergency, such as a fire, connect the pump through an independent service. Simply run wires to the pump house directly from the meter pole, ahead of circuit breakers or main fuses. Mount a weatherproof circuit breaker (or fused-disconnect switch) on the pole.

Make generous use of yard lights and locate one at the meter pole also. Use 3-way switches for control between the house and barn. Special lights are available for farmyards, but you must check the local code before connecting the lights to the wiring system.

13.9   *Lightning Arresters*

In isolated areas, lightning arrestors are a very important part of any electrical wiring system.

Even though lightning may not strike wires directly, a nearby strike may induce extremely high voltages in the wires. This can cause damage not only to the wires themselves, but to appliances and other equipment attached to the wiring system.

When a lightning arrester is installed as part of a properly grounded wiring system, the possibility of lightning damage is reduced significantly.

A lightning arrestor has three leads; one white and two black. The white lead is connected to the grounded neutral and the two black wires to the ungrounded conductors. Mount one lightning arrester at the meter pole; for improved safety mount another at each service switch to a building. The arrester "neck" is inserted through a knockout and installation is simple.

EXERCISES

13-1/   Why is an underground system recommended on a farm?

13-2/   Where does a farm service-drop terminate?

13-3/   Describe a meter pole physically.

13-4/   Are separate meters required at each building where a service entrance is employed?

13-5/   In what buildings is a ground required?

13-6/   State the NEC requirements for each of the following types of made-electrodes; concrete-encased, plate, pipe, and rod.

13-7/   How must made-electrodes be installed when rock bottom is encountered at a depth of less than 4 feet?

13-8/   What is the highest resistance-to-ground permitted by the NEC?

13-9/   Using an ammeter, describe one method for determining the resistance-to-ground.

13-10/   Give three methods by which the resistance-to-ground can be decreased.

13-11/   In connection with a meter pole, what is meant by the terms "single stack" and "double stack"?

13-12/   Where is a main switch or circuit breaker installed on a meter pole? How is it connected to the meter?

13-13/   Give the formulas for determining the minimum capacity of service conductors and service equipment at the main point of delivery to farms, including dwellings.

13-14/   A typical farm has four major loads: dairy barn, 65 A, chicken house, 26 A, storage and repair, 52 A, and residence, 60 A. What is the total demand?

13-15/   The largest motor on a farm draws 30 A. In computing the load, what current is used for the motor?

13-16/   Describe a current transformer and how it looks.

13-17/   What is a shorting switch necessary with a current transformer?

13-18/   For best protection against lightning, how is the neutral wire routed at the meter pole?

13-19/   Describe a typical ground rod installation at the foot of a meter pole.

13-20/   What are the switch requirements in all farm buildings, excluding the residence?

13-21/   Discuss the overcurrent protection requirements when buildings are fed current from other buildings.

13-22/   What type of cable is generally employed in farm electrical systems? Why?

13-23/   What types of cable must be used in buildings housing livestock?

13-24/ Describe generally how you would install lighting circuits in farm buildings other than the residence.

13-25/ What precautions should be taken to ensure operation of the water pump at all times?

13-26/ Why is a lightning arrester an essential part of a farm electrical system?

Index

G

T

FALL 77

INVENTORY 1983